Springer Series in Reliability Engineering

Series Editor

Professor Hoang Pham
Department of Industrial Engineering
Rutgers
The State University of New Jersey
96 Frelinghuysen Road
Piscataway, NJ 08854-8018
USA

Other titles in this series

Simona Salicone

Measurement Uncertainty

An Approach via the Mathematical Theory of Evidence

With 128 Figures

 Springer

Simona Salicone
Dipartimento di Elettrotecnica
Politecnico di Milano
Piazza Leonardo da Vinci 32
20133 Milano - Italy
simona.salicone@polimi.it

Mathematics Subject Classifications (2000): 94D05, 03B52, 03E75, 28E10

ISBN-13: 978-1-4419-4034-6

e-ISBN-10: 0-387-46328-3
e-ISBN-13: 978-0-387-46328-5

Printed on acid-free paper.

9 8 7 6 5 4 3 2 1

springer.com

(KeS/SB)

Contents

Preface

It is widely recognized, by the scientific and technical community that measurements are the bridge between the empiric world and that of the abstract concepts and knowledge. In fact, measurements provide us the quantitative knowledge about things and phenomena.

It is also widely recognized that the measurement result is capable of providing only incomplete information about the actual value of the measurand, that is, the quantity being measured. Therefore, a measurement result becomes useful, in any practical situation, only if a way is defined for estimating how incomplete is this information.

The more recent development of measurement science has identified in the *uncertainty* concept the most suitable way to quantify how incomplete is the information provided by a measurement result. However, the problem of how to represent a measurement result together with its uncertainty and propagate measurement uncertainty is still an open topic in the field of metrology, despite many contributions that have been published in the literature over the years.

Many problems are in fact still unsolved, starting from the identification of the best mathematical approach for representing incomplete knowledge. Currently, measurement uncertainty is treated in a purely probabilistic way, because the Theory of Probability has been considered the only available mathematical theory capable of handling incomplete information. However, this approach has the main drawback of requiring full compensation of any systematic effect that affects the measurement process. However, especially in many practical application, the identification and compensation of all systematic effects is not always possible or cost effective.

The second big unsolved problem is the definition of decision-making rules in the presence of measurement uncertainty. This problem is relevant nowadays in the measurement field for at least two reasons. The final aim of a measurement process is in fact always to take a decision. Moreover, decisions are sometimes required within the measurement process itself, because of the great diffusion of measurement instruments based on digital signal processing. In fact, these instruments allow the implementation of complex algorithms, which may contain also *if... then... else...* structures, which require a decision to be made.

The above considerations, although not exhaustive, show clearly the limitations of the currently adopted approach to the representation of measurement results and the mathematics used to handle and process incomplete information.

This book is aimed at presenting a more general approach, based on the Theory of Evidence, which encompasses the currently employed one as a particular case, and offers an effective solution to the problems that are still open.

Chapter 1 proposes a historical background of the theory of uncertainty.

Chapter 2 presents the possibility theory and the fuzzy variables and shows how these last ones are suitable variables for the representation and propagation of measurement uncertainty, when this is due to systematic contributions.

Chapter 3 presents the mathematical Theory of Evidence, defined by Shafer in 1976. The importance of this theory is that it encompasses, as particular cases, the probability theory and the possibility theory. Hence, it can be seen as the suitable theory within which variables that represent all kinds of uncertainty contributions to measurement uncertainty can be defined: the random-fuzzy variables (RFVs), which are defined in Chapter 4.

All successive chapters deal with RFVs. Chapter 5 shows how RFVs are built from the available information. Chapter 6 shows some useful fuzzy operators and, in Chapter 7, the mathematics of the RFVs is discussed. Chapter 8 shows an easy way to represent RFVs from a numerical point of view. Finally, Chapter 9 defines the decision-making rules with RFVs.

Simona Salicone
Milan, 2006

1

Uncertainty in Measurement

1.1 Introduction

In 1889, the famous physicist Lord Kelvin stated: 'When you can measure what you are speaking about, and express it in numbers, you know something about it; otherwise your knowledge is of a meager and unsatisfactory kind.'

This famous statement clearly assesses the importance of measurement and measurement science. The word *measurement* refers both to the process that is meant to provide the *number* mentioned by Lord Kelvin and to the *number* itself. This *number*, which is nothing less than the result of the measurement, is of great importance. Let us think about our everyday life. Everything accessible to our knowledge has an associated number: the weight of potatoes, the length of a table, the height of a tree, and so on. The associated numbers allow us to understand and, from the philosophical standpoint, to know. All these numbers come from measurement processes, whatever they are: from rough estimates to sophisticated procedures.

Hence, measurement results provide information; they *are* information, and therefore, they *are* knowledge. The crucial point is as follows: Does the measurement process provide *complete* knowledge about the measurand, that is, the object of measurement? Or, equivalently, which is the quality of our measurement result?

Let us try to answer the two questions. Let us suppose two children want to know 'how big is their dog.' In order to establish how big it is, they try to measure its length, from the nose to the end of the tail. The two children measure the dog one at the time, independently, each one with his own measuring tape, and achieve two different results: 110 cm and 108 cm. Is one of them wrong? Are they both wrong? Or perhaps are they both right? This last possibility could appear strange, but it is plausible in the frame of the theory of measurement, which assesses that the result of a measurement process is not only related to the object of measurement but also to the more complex system constituted by the object, the employed instruments and the environment, which can also include some human factors. In other words, variations in repeated observations are assumed to arise from not being able to hold completely constant each *influence quantity* that can affect the measurement result [ISO93]. This means that, due to the different influence quantities, including the dog's movements and the difficulty in making the measurement tape following the same path, the two children of the above example are likely to obtain two different results. Obviously, since the dog cannot be both 110 cm and 108 cm long, it must be concluded that the effect of these influence quantities has caused the deviation between the measured values and the actual length of the dog. In the following discussion, we try to understand how this deviation can be dealt with.

1.2 The Theory of Error

The Theory of Error was proposed by K. F. Gauss at the beginning of the nineteenth century, and it represents the first attempt to quantify how complete is the information provided by the measurement result. Any physical quantity is assumed to have its own *true value*, and the experimental variability of the measurement results is explained as deriving from the introduction of errors: *'while analyzing the meaning of the measurement results he had obtained, the experimenter tries to guess the true value, that is, the value that the best achievable instrument would have produced'* [M03].

In measurement science, the true value of a quantity X is defined as the value perfectly consistent with the definition of the quantity itself, and the relative measurement error of the quantity is traditionally defined by

$$e = \frac{X_m - X_t}{X_t} \tag{1.1}$$

where X_m is the measurement result of quantity X and X_t is its true value.

Moreover, errors are traditionally divided into two general classes: random and systematic.

Random errors occur with different values and signs in each different measurement (of the same quantity, with the same reference conditions). However, if the measurement of the same quantity is repeated a sufficient number of times N, the mean of these errors tends to zero and the mean of the measurement results tends to the true value of the quantity. Random errors presumably arise from unpredictable or stochastic temporal and spatial variations of the influence quantities.

The random error of a measurement result cannot be compensated by applying a correction, but it can be usefully reduced by increasing the number of observations. In fact, given N different measurements $X_1, X_2, \ldots, X_N$ (in the same conditions) of quantity X, they can be considered as N statistical variables coming from the same population, and it is therefore possible to estimate their mean and standard deviation with respect to the true value X_t:

$$\overline{X} = \frac{1}{N} \sum_{n=1}^{N} X_n \qquad (1.2)$$

$$\sigma^2 = \frac{1}{N-1} \sum_{n=1}^{N} (X_n - X_t)^2$$

The mean value expressed by Eq. (1.2) could also not coincide with any of the measured values, but it can be considered a better estimate of quantity X. In fact, if $N \to \infty$, certainly $\overline{X} \to X_t$. But also if $N \leq \infty$, $\overline{X}$ represents the best estimate of quantity X, because its associated standard deviation is $\sqrt{N}$ times lower than that associated with the single measurement result:

$$\overline{\sigma}^2 = \frac{\sigma^2}{N}$$

On the other hand, systematic errors have the following characteristic: When the measurement of a certain quantity is repeated, following the same measurement process, with the same measurement instruments and the same reference conditions, they always occur with the same value and sign.

From a strict theoretical point of view, systematic errors, differently from random ones, can be fully compensated, and their effect can be eliminated. This result may happen, however, only if the systematic errors are totally recognized. On the contrary, from a practical point of view, their effects can be generally only reduced, and not eliminated completely.

If more systematic contributions are present in the measurement process and they can be determined separately, it is possible to combine them and find the final measurement error that affects the measured quantity. In particular, if measurand X depends on quantities $A_1, A_2, \ldots, A_n$, the following applies[1]:

$$X_m = f(A_1, A_2, \ldots, A_n)$$

$$e = \frac{X_m - X_t}{X_t} \cong \frac{dX}{X} = \frac{\sum \frac{df}{dA_k} dA_k}{f(A_1, A_2, \ldots, A_n)} = \sum \alpha_k \frac{dA_k}{A_k}$$

where

$$\alpha_k = \frac{df}{dA_k} \frac{A_k}{f}$$

Therefore, the final relative error is a linear combination of the relative errors of $A_1, A_2, \ldots, A_n$.

According to the theory of errors, if the errors associated with N different measurements of the same quantity (performed in the same conditions) have different values but have a mean different from zero, it can be concluded

---

[1] The approximation in the formula is acceptable because, generally, the measurament error is much smaller than the measurand and therefore, it approximates the differential dX.

that the mean of the errors is due to a systematic effect and the residual differences are due to random effects. If, for instance, the same example of the two children and the dog, shown in the previous section, is reconsidered, the following can apply. Let us consider two situations: In the first one, let us suppose the dog is 109 cm long; in the second one, let us suppose the dog is 109.5 cm long. In both cases, none of the children has found the true value of the length of the dog; hence, both of them did some measurement errors. These errors can be due to both systematic (the use of different measuring tapes) and random contributions (the inevitable movements of the dog, the different path followed by the measuring tape from the nose of the dog to its tail and so on). In particular, in the first case, the relative errors done by the two children are $e_1 = +1/109$ and $e_2 = -1/109$, respectively. As the mean of the errors $e_m = (e_1 + e_2)/2$ is zero, the Theory of Error allows us to state that they are due only to random contributions. In the second case, on the other hand, the errors are $e_1 = +0.5/109.5$ and $e_2 = -1.5/109.5$, respectively, and their mean is $e_m = -0.5/109.5$. Thus, the Theory of Error allows us to state that the error e_m is systematic and the residual error $\pm 1/109$ is random.

Considering the above example, it can be noted that, in order to apply the Theory of Error and find these results, we had to suppose the knowledge of the true value of the length of the dog. Then, a very simple but important question arises: If any measurement result is affected by errors, how is it possible to know the true value of the measured quantity?

This simple question shows the intrinsic limit of the Theory of Error: The true value, which represents the basis of this theory and shows up in Eq. (1.1), is an ideal concept and cannot ever be known exactly. It follows that the measurement error defined by Eq. (1.1) can never be evaluated. Thus, the Theory of Errors fails.

However, at this point, an answer to the crucial question asked in the previous section has been already found: The knowledge provided by a measurement process, that is the measurement result, is always incomplete. But this answer brings us to another question: If the measurement result provides incomplete information about the measurand, how can this measurement result be qualified? Or, in other words, how can this incomplete knowledge be represented?

1.3 The Theory of Uncertainty

At the end of the twentieth century, the Theory of Error was replaced by the Theory of Uncertainty. As also reported in [ISO93], the uncertainty is an attribute of the measurement result, which reflects the lack of exact knowledge of the value of the measurand.

The changeover from error to uncertainty is officially set by the publication, in 1992, of the IEC-ISO "Guide to the Expression of Uncertainty in Measurement" (GUM) [ISO93]. The old concept of error disappears, because it is an ideal concept and it is impossible to evaluate, and the new concept of uncertainty is introduced: '*a parameter, associated with the result of a measurement, that characterizes the dispersion of the values that could reasonably be attributed to the measurand.*' According to the GUM, this parameter is statistically determined. In fact, because '*it is assumed that the result of a measurement has been corrected for all recognized significant systematic effects,*' the result of a measurement is only affected by random effects and therefore can be mathematically represented, in the frame of the probability theory, by a probability density function. Hence, the *standard uncertainty* associated with the result of a measurement is the standard deviation of such a probability density function [ISO93].

The problem now becomes how to estimate such a probability density function so that the standard uncertainty can also be estimated.

The most immediate way is the experimental one: The measurement procedure is repeated several times, and the standard deviation of the obtained measurement results is computed. However, this procedure takes time and cannot be always followed, mainly for practical and economical reasons.

On the other hand, it is also recognized that the probability density function associated with the measurement results can be often assumed starting from a priori knowledge, such as a calibration certificate, the experience of the operator, and so on.

These two ways for estimating the probability density function and the appertaining standard deviation are equally reliable and are both considered by the GUM, which calls them *type A evaluation* and *type B evaluation* of uncertainty.

Furthermore, type A evaluation of the standard uncertainty is obtained by means of a statistical analysis of a series of observations. Given N different measurements $X_1, X_2, \ldots, X_N$ obtained under the same conditions of measurement, the best available estimate of the expected value of quantity X is the arithmetic mean of the N observations $(\overline{X})$ and the best estimate of the variance of the mean is [P91]:

$$s^2(\overline{X}) = \frac{1}{N\,(N-1)} \sum_{n=1}^{N} (X_n - \bar{X})^2$$

The positive square root of the variance of the mean is the experimental standard deviation of the mean, and it quantifies how well $\overline{X}$ estimates the expected value of X; therefore, it may be used to measure the standard uncertainty of $\overline{X}$, generally named u, and qualify the measurement result itself.

Type B evaluation of the standard uncertainty is obtained by means of judgment using all relevant information on the possible variability of quantity X. The available information may include previous measurement data, experience or general knowledge, manufacturer's specifications, data provided in calibration certificates, and so on. For instance, if the available information leads to the conclusion that the measurement result distributes according to a rectangular distribution, with base $(a_+ - a_-) = 2a$, then the expected value of X is $(a_- + a_+)/2$ and its standard uncertainty is $s(\overline{X}) = a/\sqrt{3}$. If information is available that the measurement result distributes according to a symmetric trapezoidal distribution having a base of width $(a_+ - a_-) = 2a$ and a top of width $2a\beta$, where $0 \leq \beta \leq 1$, then the standard uncertainty becomes $s(\overline{X}) = \sqrt{a^2 (1 + \beta^2)/6}$. When $\beta = 1$, the trapezoidal distribution becomes a rectangular distribution and the already mentioned result is obtained; when $\beta = 0$, the trapezoidal distribution becomes a triangular distribution and the standard deviation is given by $s(\overline{X}) = a/\sqrt{6}$ [P91].

The above definitions apply to uncertainty estimation in direct measurements, that is, in those measurement processes in which the measurand can be measured directly. However, several quantities are measured indirectly by suitably processing the results of other measurements. It is therefore important that the uncertainty concept, similarly to the error concept, is also applied to these indirect measurements, and that the uncertainty associated with the final measurement result is obtained from the uncertainty values associated with each processed measured value.

When the standard uncertainty is considered, a *combined standard uncertainty* can be associated with the final measurement result as follows.

Let Y be the measurand, $X_1, X_2, \ldots, X_n$, the measurable quantities on which the measurand depends, and $Y = f(X_1, X_2, \ldots, X_n)$, their mathematical relationship. The expected value y of quantity Y is determined by the expectations $x_1, x_2, \ldots, x_n$ of quantities $X_1, X_2, \ldots, X_n$. Moreover, if f is a linear function, the combined standard uncertainty $u_c(y)$ of the measurement result y can be obtained as follows [ISO93]:

$$u_c(y) = \sqrt{\sum_{i=1}^{n} \left(\frac{\partial f}{\partial x_i}\right)^2 u^2(x_i) + 2\sum_{i=1}^{n-1} \sum_{j=i+1}^{n} \frac{\partial f}{\partial x_i} \frac{\partial f}{\partial x_j} u(x_i, x_j)} \qquad (1.3)$$

where $\frac{\partial f}{\partial x_i}$, called the *sensitivity coefficient*, is the partial derivative of function f with respect to quantity x_i and evaluated in $x_1, x_2, \ldots, x_n$; $u(x_i)$ is the standard uncertainty associated with quantity x_i; $u(x_i, x_j)$ is the estimated covariance of x_i and x_j. Obviously, it is

$$u(x_i, x_j) = u(x_j, x_i)$$

The covariance between x_i and x_j can be determined starting from the standard uncertainties associated with quantities x_i and x_j and the correlation coefficient $\varphi(x_i, x_j)$, which expresses the degree of correlation between x_i and x_j:

$$u(x_i, x_j) = u(x_i) \, u(x_j) \, \varphi(x_i, x_j)$$

The correlation coefficient satisfies the following properties:

$-1 \leq \varphi(x_i, x_j) \leq 1$;

$\varphi(x_i, x_j) = \varphi(x_j, x_i)$;

$\varphi(x_i, x_j) = 0$ when quantities x_i and x_j are uncorrelated.

Equation (1.3) is referred to as the standard uncertainty propagation law [ISO93].

If f is not a linear function, Eq. (1.3) represents the first-order Taylor-series approximation of $Y = f(X_1, X_2, \ldots, X_n)$. However, it can be still considered in the evaluation of $u_c(y)$ in most practical situations. In fact, in the measurement practice, the value of the measurement uncertainty is generally small with respect to the measured value and hence determines small variations of the measurand. This means that the linearity condition of f is almost always locally verified, near the measurement point, and Eq. (1.3) remains valid.

On the contrary, when the nonlinearity of f is significant, higher order terms in the Taylor series must be considered in Eq. (1.3).

It can be now recognized that the standard uncertainty (and the combined standard uncertainty, when indirect measurements are considered) meets the primary goal of characterizing 'the dispersion of the values that could reasonably be attributed to the measurand' [ISO93], without any need to refer to the unknown and unknowable true value of the measurand, provided that the sole causes for this dispersion are random.

However, there is still one problem left, whose solution cannot be found directly in terms of standard uncertainty. As stated by the GUM, 'in many industrial and commercial applications, as well as in the areas of health and safety, it is often necessary to provide an interval about the measurement result within which the values that can reasonably be attributed to the quantity subject to measurement may be expected to lie with a high level of confidence. Thus the ideal method for evaluating and expressing measurement uncertainty should be capable of readily providing such a confidence interval, in particular, one that corresponds in a realistic way with the required level of confidence' [ISO93].

The confidence interval of a measurement result y is generally given as an interval centered on the result $(y \pm U)$ and can be found only if the probability distribution of the result is known, once the required level of confidence has been fixed. In fact, if $g(y)$ is the probability distribution of Y, then its mean value and standard deviation are given by [P91]:

$$\overline{y} = \int_{-\infty}^{+\infty} y\, g(y)\, dy$$

$$u(y) = \sigma_{g(y)} = \sqrt{\int_{-\infty}^{+\infty} (y - \overline{y})^2\, g(y)\, dy}$$

If now p is the required level of confidence, the associated confidence interval is found by solving the following integral:

$$\int_{\overline{y}-U}^{\overline{y}+U} g(y)\, dy = p \tag{1.4}$$

The semi-width U of the confidence interval, which is called *expanded uncertainty* in [ISO93], is generally given as a multiple of the standard uncertainty:

$$U = k_p \, u(y)$$

This means that the problem of determining a confidence interval is given by the difficulty in the determination of k_p. Of course, the multiplying factor k_p, called in [ISO93] the *coverage factor*, strictly depends on the probability distribution $g(y)$ and the level of confidence p.

For instance, if $g(y)$ is a normal distribution, the values of the coverage factors correspondent to the most commonly used confidence intervals are as follows:

- $k_p = 1$ for $p = 68.27\%$
- $k_p = 1.645$ for $p = 90\%$
- $k_p = 1.960$ for $p = 95\%$
- $k_p = 2$ for $p = 95.45\%$
- $k_p = 2.576$ for $p = 99\%$
- $k_p = 3$ for $p = 99.73\%$

These values are tabulated in [ISO93] and in each book of probability and statistics.

If the measured quantity Y can be measured directly, then the probability distribution of the result can be found by following a type A or a type B evaluation. This task is sometimes difficult, but it generally leads to a solution. On the other hand, when the measured quantity Y can be only measured indirectly, the determination of the probability distribution of the result becomes more difficult.

If the probability distributions of the input quantities $X_1, X_2, \ldots, X_n$ are known and relationship f between Y and its input quantities is a linear function, then the probability distribution of Y may be readily obtained by a series of convolutions. But if relationship f is nonlinear and a first-order Taylor expansion is not an acceptable approximation, the distribution of Y cannot be simply expressed as a convolution. In these situations, numerical methods (such as the Monte Carlo one) are required [ISO04].

However, the convolution of probability distributions is complex and the Monte Carlo simulation has a high computational burden.

Sometimes, however, it is not necessary to handle all probability distributions, and the knowledge of the correspondent standard deviations is enough to determine the appropriate coverage factor associated with the fixed level of confidence. This happens whenever the assumptions of the Central Limit Theorem, which is one of the most remarkable results of the probability theory, are met.

Theorem 1.1 (The Central Limit Theorem). *Let $X_1, X_2, \ldots, X_n$ be independent random variables sharing the same probability distribution D. As-*

sume that both the expected value μ and the standard deviation σ exist and are finite.

Consider the sum $S_n = X_1 + X_2 + \cdots + X_n$. Then the expected value of S_n is $n\mu$ and its standard deviation is $\sigma\sqrt{n}$.

Furthermore, as n approaches ∞, the distribution of S_n approaches the normal distribution $N\left(n\mu, n\sigma^2\right)$.

Under the assumptions of this theorem, it is known that the result of the indirect measurement follows a specific normal distribution; hence, no calculations are required, and the coverage factors can be found on tables.

Moreover, it can be demonstrated that the thesis of the Central Limit Theorem is still verified when the hypothesis are less strict. In fact, a generalization of the Central Limit Theorem exists that does not require identical distributions of the initial random variables. Of course, this generalized version of the Central Limit Theorem requires some additional conditions, which ensure that none of the variables exerts a much larger influence than the others.

Under this assumption, the Central Limit Theorem assures that the random variable representing the final result $Y = f(X_1, X_2, \ldots, X_n)$ approximates a normal distribution. Mean value and standard deviation of this normal distribution are respectively given by $y = f(x_1, x_2, \ldots, x_n)$ and by Eq. (1.3).

Moreover, the approximation is better as the number of input quantities increases; the convergence is more rapid the closer the values of the variances are to each other; and the closer the distributions X_i are to normal ones, the fewer X_i are required to yield a normal distribution for S_n. To verify how this statement holds in practice, the following example is given.

Let us consider the sum of rectangular distributions, which are a significant example of a non-normal distribution. As stated, the probability density function of the sum is mathematically obtained by convoluting the initial distributions. Figure 1.1 shows the convolution (dot line) of four rectangular distributions having expected value $\mu = 10$ and width 0.2, which means a standard deviation $\sigma = 0.1/\sqrt{3}$. The validity of the Central Limit Theorem is readily proven, because this convolution is approximately normal (solid line). Therefore, the convolution of only four rectangular distributions of equal width is approximately normal.

Figure 1.2 shows the convolution (dot line) of four rectangular distributions having again expected value $\mu = 10$ but with different widths (0.16, 0.08, 0.2 and 0.4, respectively). It can be noted that, in this case, four rectangular distributions are not enough to approximate a normal distribution (solid line). On the other hand, seven rectangular distributions could be enough, as shown in Fig. 1.3. In this case, the distributions have once again an expected value $\mu = 10$ and widths 0.16, 0.08, 0.2, 0.4, 0.1, 0.02, and 0.3 respectively, which differ also of one order of magnitude.

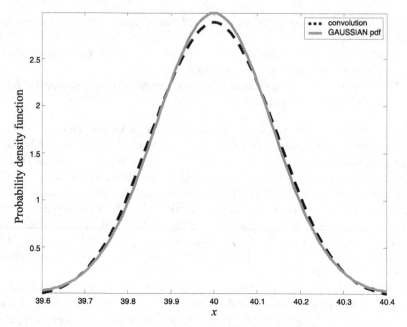

Fig. 1.1. Convolution of four rectangular distributions having same mean and variance.

This example confirms, in an intuitive way, that the values that can be attributed to the measurand, when obtained by means of an indirect measurement process, distribute according to a fairly normal distribution, provided that the hypothesis of the Central Limit Theorem is satisfied (at least at a first approximation). Therefore, the combined standard uncertainty provided by Eq. (1.3) is the standard deviation of such a distribution and any required confidence interval with assigned confidence level can be readily obtained by multiplying this standard deviation by a coverage factor typical for the normal distribution.

1.4 Toward a more modern and comprehensive approach

The approach based on the uncertainty concept is the first serious attempt to quantify the incompleteness of the available information without the need to refer to the true value of the measurand.

However, this approach requires that a suitable mathematics is used to represent the incomplete information and combine different contributions with each other. Since the Theory of Uncertainty was proposed, the probability theory has been identified as the deputy mathematical tool for handling uncertainties. The reason for this choice probably depends on the fact that,

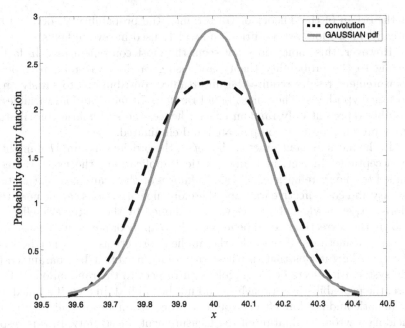

Fig. 1.2. Convolution of four rectangular distributions having different widths.

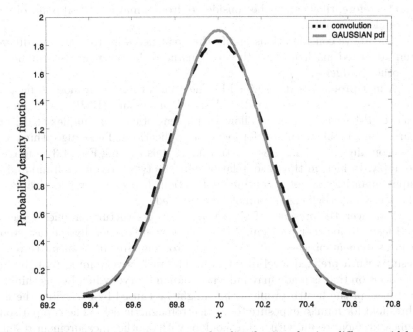

Fig. 1.3. Convolution of seven rectangular distributions having different widths.

at the origin of the Theory of Uncertainty, the probability theory had been already known for two centuries, and widely used in several fields.

However, this choice does not seem the most convenient one. In fact, referring to the probability theory and using random variables to represent measurement results require the uncertainty contributions to satisfy an important hypothesis: They must be all random. In fact, random variables can suitably represent only random effects; hence, the systematic contributions, when present, must be compensated and eliminated.

In the measurement practice, however, the previous assumption might not be acceptable. In fact, in many practical applications, the exact values assumed by the systematic effects are unknown or they cannot be extimated in an easy and cost-effective way, and therefore, no corrections can be performed. Moreover, even when the exact values assumed by the systematic effects are known, the corrections could be impossible to apply or not convenient in terms of costs; sometimes, if possible, the applied corrections could not be totally effective. This means that, in these situations, it cannot be guaranteed that the systematic effects are negligible with respect to the random ones. Therefore, the probabilistic approach should not be applied. In fact, if applied, both systematic and random effects would be processed in a probabilistic way, thus yielding a wrong evaluation of the measurement uncertainty. In this respect, let us consider that systematic effects do not propagate in a statistical way, and therefore, they cannot be considered, from a mathematical point of view, as random variables.

The above considerations justify the great necessity to refer to a different and advanced mathematical theory. This need, however, is also emphasized by other problems.

A first problem is determined by the great evolution of modern measurement instruments based on digital signal processing (DSP) techniques. In fact, DSP-based instruments allow the implementation of complex algorithms, barely imaginable until now for low-cost applications. These algorithms could also contain $if\dots then\dots else\dots$ structures, thus making Eq. (1.3) impossible to apply. In fact, in this case, relationship f between the measurand and the input quantities is not a continuous function; hence, f is not derivable and the sensitivity coefficients cannot be evaluated.

Moreover, the presence of $if\dots then\dots else\dots$ structures implies a decision and leads to another problem. Taking a decision is generally the last step of a measurement process and requires one to compare the final measurement result with a prefixed level or interval. The task is very important, because the decision that is made may influence human everyday life. Let us think, for instance, of the health problem that might be caused by exceeding the fixed threshold for human exposure to an electromagnetic field. The comparison of a measurement result with a threshold, or with another measurement result, is quite troublesome when the measurement uncertainty is taken into account, because it requires the comparison of an interval with a number, or with another interval, which sometimes could overlap.

In modern measurement systems, which execute a complex measurement algorithm, due to the possible presence of *if. . . then. . . else. . .* structures, decisions could be required not only at the end of the process, but also within the measurement algorithm. This means that the importance of taking correct decisions becomes even more important, because the final measurement result, on which the final decision is based, also depends on the intermediate decisions.

The solution of this problem is not immediate when the probabilistic approach is followed. Hence, once again, the necessity of a new, more general method to deal with measurement uncertainty emerges.

The aim of this book is indeed to give a contribution to the Theory of Uncertainty, trying to overcome the limits underlined above. The result of a measurement can be seen as an incomplete knowledge of the measurand itself; the knowledge is incomplete due to the presence of some influence quantities, which are generally divided into two main classes: quantities that affect the measurement result in a systematic way and those that affect the measurement result in a random way.

The first question to be answered is as follows: As systematic contributions, even if unknown, could be present in the measurement process, how can they be mathematically represented? The answer to this question is given in Chapter 2: Considering the way systematic contributions combine themselves, their effect on the measurement result can be framed within the possibility theory and can be represented by fuzzy variables.

After having shown that fuzzy variables can be used to take into account the effects of systematic contributions, it can be concluded that random variables, which are the natural variables of the probability theory, can correctly represent, from a mathematical point of view, random contributions to the measurement uncertainty; fuzzy variables, which are the natural variables of the possibility theory, can correctly represent, from a mathematical point of view, systematic contributions to the measurement uncertainty.

Probability and possibility theories are two distinct mathematical theories, but they have a common root, which is represented by the mathematical Theory of Evidence. In fact, the probability and the possibility theories are two distinct, particular cases of the more general mathematical Theory of Evidence. Therefore, in order to solve the problem to consider all kinds of contributions to the measurement uncertainty, and not only the random ones (if the probability theory is considered), or the systematic ones (if the possibility theory is considered), this more general theory should be referred as well. Chapter 3 is entirely dedicated to the mathematical Theory of Evidence, defined by Shafer in 1976.

Chapter 4 comes back to the original problem of measurement uncertainty and deals with random–fuzzy variables (RFVs), which are defined within the Theory of Evidence and can be employed to represent, in a unique mathematical object, all kinds of contributions to the measurement uncertainty.

The successive chapters deal again with RFVs. In particular, Chapter 5 deals with the problem of building an RFV starting from the available information. When the cause of incomplete knowledge is due to systematic effects or total ignorance, the construction is immediate. On the other hand, when the cause of incomplete knowledge is due to random effects, the construction requires the definition of a suitable probability–possibility transformation.

In Chapter 6, different fuzzy operators are defined, whereas in Chapter 7, the mathematics of RFVs is defined, taking into account the real behavior of the different contributions to uncertainty.

Chapter 8 briefly shows how RFVs can be represented from a numerical point of view, whereas, finally, Chapter 9 deals with the decision-making rules, which is a relevant problem in the measurement field. It is shown that decision making is immediate when RFVs are considered.

2

Fuzzy Variables and Measurement Uncertainty

Chapter 1 has presented a short survey of the basic concepts of the measurement theory. In particular, it has shown that the result of a measurement represents incomplete knowledge of the measurand and this knowledge can be usefully employed only if its 'incompleteness' can be somehow estimated and quantified. It has also been shown that this approach requires a suitable mathematics for handling, from the quantitative point of view, incomplete knowledge. Current practice refers mainly to the probability theory, because it is the best known and most assessed mathematical theory that treats incomplete knowledge.

However, the probability theory deals only with that particular kind of incomplete knowledge originated by random effects. As seen in the previous chapter, this implies that all other recognized significant effects, including the systematic ones, are fully compensated.

A systematic effect affects the measurement process always with the same value and sign. If this value was exactly known, it could be treated as an *error* and the measurement result could be corrected by compensating it in the measurement procedure. Of course, this situation does not generally occur. The normal situation is that the presence of a systematic contribution is recognized, but its exact value is not known, even if it is possible to locate it within a closed interval of $\Re$, that is, from a mathematical point of view, an *interval of confidence*. By definition, a systematic contribution always takes the same value, although unknown, within this estimated interval. This means that each value of the interval does not have the same probability to occur, but it has, in absence of further evidence, the same possibility to occur, because no value is preferable to the others.

This situation can be also used to represent a more general case. In many practical situations, there is evidence that some unknown effect is affecting the measurement result. The only available information shows that the contribution related to this unknown effect falls, with a given level of confidence, within a given interval, but it is not known whereabouts. Moreover, nothing else can be assumed, not even if the effect is a systematic or a random one. This situation is called *total ignorance*.

Even if the two kinds of contributions are represented in the same way—an interval of confidence—they should not be confused with each other. In fact, when total ignorance is considered, no other information is available; on the contrary, when a systematic contribution is considered, an important piece of information is available, that is, the systematic behavior of the contribution itself. This additional information should be carefully taken into account when modeling the measurement process.

Let us consider the following simple example. Let m_1 and m_2 be two measurement results for the same quantity, for instance, the weight of an object obtained with the method of the double weighing with a weighbridge. The final measurement result is supposed to be obtained by the arithmetic mean of m_1 and m_2. Let us suppose that no random contributions are present, but only one contribution of a different nature, whose possible values fall within interval ± 100 g. Two cases can be considered:

1. No additional information is available for this uncertainty contribution, so that it must be classified as total ignorance. In this case the following applies:

$$r = \frac{(m_1 \pm 100 \ g) + (m_2 \pm 100 \ g)}{2} = \frac{(m_1 + m_2)}{2} \pm 100 \ g$$

    and the measurement uncertainty associated with the result is the same as the initial one.

2. Additional information is available showing that the uncertainty contribution is due to the different length of the two beams of the weighbridge. In this case, thanks to the available information about this uncertainty contribution, it is known that it affects m_1 and m_2 with the same absolute value, even if not known, and opposite signs (in fact, let us remember that m_1 and m_2 are obtained by placing the object to be weighed on the two different plates of the weighbridge). Therefore, the contribution can be classified as systematic and the following applies:

$$r = \frac{(m_1 \pm 100 \ g) + (m_2 \mp 100 \ g)}{2} = \frac{(m_1 + m_2)}{2}$$

    and the final measurement result has zero associated uncertainty because, in the arithmetic mean, the considered contribution is compensated.

This simple example shows the importance of using all available information while modeling a measurement process. In fact, if the additional information about the reason for the systematic behavior was not taken into account, the measurement uncertainty of the final result would have been overestimated.

Hence, it can be stated, also from a mathematical point of view, that a systematic contribution is a particular case of total ignorance, where additional information is added. If this additional information is given, together with the reasons of the presence of the systematic contribution itself, then it can sometimes be used to suitably and correctly propagate the uncertainty through the measurement process, as in the case of the double weighing with the weighbridge.

The aim of this chapter is to find a mathematical object able to represent total ignorance and its particular cases.

When the probability theory is considered to handle incomplete knowledge, total ignorance is represented by a random variable with a uniform probability distribution over the estimated confidence interval. However, this assumption is inconsistent with the concept of *total* ignorance.

In fact, let us first consider the general case of total ignorance. In this case, by definition, no assumptions can be made about the actual probability distribution, and all probability distributions are, in theory, possible. Hence, assuming a uniform probability distribution means to arbitrarily add information that is not available.

Let us now consider the particular case of a systematic effect. In this case, by definition, only one unknown value has 100% probability to occur, whereas all others have null probability. Hence, assuming a uniform probability, which implies that all values within the given interval have the same probability to occur, leads to a bad interpretation of the available information.

Of course, the assumption of whichever other probability distribution brings one to the same considerations. Hence, it can be concluded that the probability theory and the random variables are not able to represent incomplete knowledge, whenever this is due to uncertainty effects that are not explicitly random. Therefore, different mathematical variables should be considered for this aim.

In the second half of the twentieth century, fuzzy variables have been introduced in order to represent incomplete knowledge. This approach is less known than the statistical one, and it is completely lacking in the current standards; therefore, the whole chapter is dedicated to this approach.

2.1 Definition of fuzzy variables

Fuzzy variables and fuzzy sets have been widely used, in the last decades, especially in the field of automatic controls, after Zadeh introduced the basic principles of fuzzy logic and approximate reasoning [Z65]-[Z73]-[Z75]-[Z78].

In the traditional mathematical approach, a variable may only belong or not belong to the set into which it is defined. The function describing the membership of such a crisp variable to its appertaining set can therefore take only the value 1, if the variable belongs to the set, or 0, if the variable does not belong to the set.

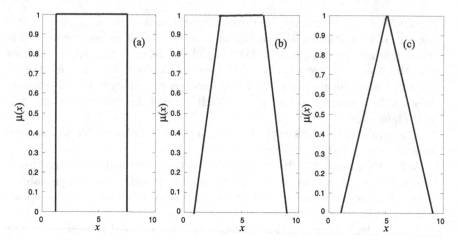

Fig. 2.1. Example of membership functions: (**a**) rectangular, (**b**) trapezoidal, and (**c**) triangular.

When fuzzy variables and fuzzy sets are considered, the function describing the membership of a variable to its appertaining set is allowed to take all values within the 0–1 interval. This means that, given the referential set $\Re$ of the real numbers, a fuzzy variable X is defined by its membership function $\mu_X(x)$, where $x \in \Re$. The membership function of a fuzzy variable satisfies the following properties [KG91]:

- $0 \le \mu_X(x) \le 1$;
- $\mu_X(x)$ is convex;
- $\mu_X(x)$ is normal (which means that at least one element x always exists for which $\mu_X(x) = 1$).

Figure 2.1 shows some examples of membership functions: The rectangular one, the trapezoidal one, and the triangular one. Of course, different shapes are allowed, which do not need to be symmetric. Moreover, membership functions that only increase or only decrease qualify as fuzzy variables [KY95] and are used to represent the concept of 'large number' or 'small number' in the context of each particular application. An example of such a membership function is given in Fig. 2.2.

The membership function $\mu_X(x)$ of a fuzzy variable can be also described in terms of α-cuts at different vertical levels α. As the membership function of a fuzzy variable ranges, by definition, between 0 and 1, its α-cuts are defined for values of α between 0 and 1.

Each α-cut, at the generic level α, is defined as

$$X_\alpha = \{x \mid \mu_X(x) \ge \alpha\} \tag{2.1}$$

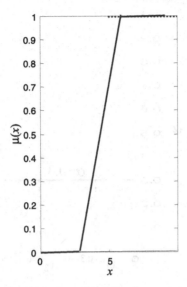

Fig. 2.2. Membership function describing the fuzzy statement 'greater than 5.'

According to Eq. (2.1), each α-cut defines an interval $[x_1^\alpha, x_2^\alpha]$, where it is always $x_1^\alpha \leq x_2^\alpha$. The equality of x_1^α and x_2^α can be reached only for $\alpha = 1$ and only if the membership function has a single peak value, for instance, in the case of the triangular membership function reported in Fig. 2.1c. Generally, x_1^α and x_2^α take finite values, but in some cases, as the one in Fig. 2.2, it could be $x_1^\alpha = -\infty$ and/or $x_2^\alpha = +\infty$. If, for any value α, it is $x_1^\alpha = x_2^\alpha$, then the fuzzy variable degenerates into a crisp variable. An example of α-cut, for level $\alpha = 0.3$, is given in Fig. 2.3.

The importance of representing a fuzzy variable in terms of its α-cuts is that the α-cuts of a fuzzy variable and the corresponding levels α can be considered as a set of intervals of confidence and associated levels of certitude. The level of certitude contains information about how certain a person is about its knowledge. If a person, for instance, remembers exactly the birthday of a friend, his knowledge is certain; but if he only remembers the month, but not the day, the certainty of his knowledge is lower. In the first case, the available knowledge can be represented by a crisp value, that is, the exact day of the year (April, 15th); in the second case, the available knowledge can be represented by a set, which contains the 30 days of April. Hence, as the level of certitude increases, the width of the corresponding confidence interval decreases.

As simply shown by the previous example, the link between the level α of certitude and the confidence interval at the same level corresponds to the natural, often implicit, mechanism of human thinking in the subjective estimation of a value for a measurement.

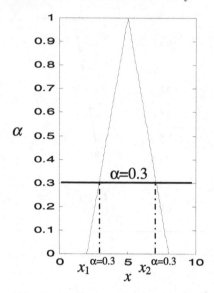

Fig. 2.3. α-cut of a triangular membership function for level $\alpha = 0.3$.

The following example shows how a fuzzy variable can be built, starting from the information, or personal idea, of one, or more human beings. Let us consider again the birthday of a friend, whose name is Sam. Tom is sure that Sam's birthday is April 15th; John is not so sure about the date, even if he remembers well that the birthday falls in April. From a mathematical point of view, the interval of confidence estimated by Tom is [April 15, April 15], whereas the interval of confidence estimated by John is [April 1, April 30]. The levels of certitude associated with the two intervals are 1 and 0, respectively. If now a fuzzy variable must be associated with the available information, the following applies.

Let us suppose the only available information is from Tom. In this case, the available information can be represented by a crisp value. In fact, in this case, full certainty is given.

Let us now suppose the only available information is from John. In this case, the available information can be represented by a rectangular fuzzy variable, like the one in Fig. 2.1a.

However, it is also possible to combine the information given by Tom and John. As the level of certitude associated with John's interval is zero and those associated with Tom's interval is one, these two intervals can be considered as the α-cuts of the fuzzy variable, which represents Sam's birthday, at levels of α zero and one, respectively. Since, as the level of certitude increases, the interval of confidence becomes narrower, the fuzzy variable could be, for instance, triangular, like the one in Fig. 2.1c.

This simple example also shows, in an intuitive way, a relationship between the level of certitude and the level of confidence, which, according to the Theory of Uncertainty, should always be associated with an interval of confidence. The level of certitude indicates how much a person is sure about a certain event. If, for instance, the example of Sam's birthday is considered again, this means that the surer a person is about the birthday's date, the smaller is the extimated range of days. On the other hand, the level of confidence indicates the probability of a certain event. Therefore, considering the same example of Sam's birthday, the smaller the given range of days, the smaller the probability that Sam's birthday falls within those dates. If only one date is given, the probability that this is exactly Sam's birthday is zero, as also shown by Eq. (1.4). In other words, although the level of certitude increases as the width of the confidence interval decreases, the opposite applies to the levels of confidence. Hence, the levels of confidence equal to one and zero are assigned to intervals [April 1, April 30] and [April 15, April 15], respectively. Moreover, intuitively, it can also be assessed that, given the confidence interval at level α, the associated level of confidence is $1 - \alpha$[1] [FS03, FGS04, FS04, FS05a, FS05b, FS05c].

The use of fuzzy variables in the context of measurement uncertainty is particularly interesting, if we remember that the most useful way to express the result of a measurement is in terms of confidence intervals and '*the ideal method for evaluating and expressing measurement uncertainty should be capable of readily providing such a confidence interval*' [ISO93]. Therefore, it can be stated that a fuzzy variable can be effectively employed to represent the result of a measurement, because it provides all available information about the result itself: the confidence intervals and the associated levels of confidence.

The good measurement practice also requires that the uncertainty of a measurement result is directly usable as a component in evaluating the uncertainty of another measurement in which the first result is used [ISO93]; that is, measurement uncertainties have to be composed among each other. Thus, arithmetic operations among fuzzy variables must be defined, in order to ensure that fuzzy variables are able to propagate measurement results and related uncertainties.

2.2 Mathematics of fuzzy variables

As shown in the previous section, a fuzzy variable A can be described in two ways: by means of its membership function $\mu_A(x)$, for $x \in \Re$; or by means of its α-cuts A_α, for $0 \leq \alpha \leq 1$. These two ways to represent a fuzzy variable are, of course, equivalently valid and contain the same information, because the α-cuts can be determined starting from the membership function and vice versa. Similarly, the mathematics of fuzzy variables can also be defined in two different ways, which refer to membership functions and α-cuts, respectively.

---

[1] This statement will be proven, in a more general context, in the next chapter. For the moment, let us rely on the given intuitive example.

As the α-cuts of a fuzzy variable are confidence intervals and the aim of a measurement process is indeed to find confidence intervals, the second approach is preferable and hence reported in this section.

Let us consider fuzzy variables with a finite support, like, for instance, the ones in Fig. 2.1. Even if these variables are only a part of the whole kind of possible fuzzy variables, as shown in the previous section, this assumption is coherent with the measurement practice, where the possible values that can be assumed by a measurand are almost always confined into a closed interval. Moreover, even when the probability distribution of the measurement results is supposed to be greater than zero in every point of $\Re$, like, for instance, a Gaussian distribution, the probability that the measurand takes values outside a suitable confidence interval is very small and it is possible to consider this last interval as the support of the fuzzy variable, which correspond to the Gaussian distribution.[2]

Under the assumption of finite support, fuzzy arithmetic can be then defined on the basis of the two following properties:

- Each fuzzy variable can fully and uniquely be represented by its α-cuts.
- The α-cuts of each fuzzy variable are closed intervals of real numbers.

These properties enable one to define arithmetic operations on fuzzy numbers in terms of arithmetic operations on their α-cuts or, in other words, arithmetic operations on closed intervals [KG91, KY95]. These operations are a topic of the *interval analysis*, a well-established area of classic mathematics. Therefore, an overview of the arithmetic operations on closed intervals is previously given.

Let $*$ denote any of the four arithmetic operations on closed intervals: addition $+$, subtraction $-$, multiplication $\times$, and division $/$. Then, a general property of all arithmetic operations on closed intervals is given by

$$[a, b] * [d, e] = \{f * g \mid a \leq f \leq b, d \leq g \leq e\} \qquad (2.2)$$

except that the division is not defined when $0 \in [d, e]$. The meaning of Eq. (2.2) is that the result of an arithmetic operation on closed intervals is again a closed interval.

This last interval is given by the values assumed by the proper operation $f * g$ between numbers f and g, taken from the original intervals. In particular, the four arithmetic operations on closed intervals are defined as follows:

$$[a, b] + [d, e] = [a + d, b + e] \qquad (2.3)$$

$$[a, b] - [d, e] = [a - e, b - d] \qquad (2.4)$$

$$[a, b] \times [d, e] = [\min(ad, ae, bd, be), \max(ad, ae, bd, be)] \qquad (2.5)$$

---

[2] The information contained in a probability distribution can also always be represented in terms of a fuzzy variable. This is possible thanks to suitable probability–possibility transformations, as will be shown in Chapter 5.

and, provided that $0 \notin [d, e]$

$$[a, b] / [d, e] = [a, b] \times [1/e, 1/d]$$
$$= [\min (a/d, a/e, b/d, b/e), \max (a/d, a/e, b/d, b/e)]$$
(2.6)

It can be noted that a real number r may also be regarded to as a special (degenerated) interval $[r, r]$. In this respect, Eqs. (2.3) to (2.6) also describe operations that involve real numbers and closed intervals. Of course, when both intervals degenerate, the standard arithmetic on real numbers is obtained.

The following examples illustrate the arithmetic operations over closed intervals, as defined by Eqs. (2.3)–(2.6):

$$[2, 5] + [1, 3] = [3, 8] \quad [0, 1] + [-6, 5] = [-6, 6]$$

$$[0, 1] - [-6, 5] = [-5, 7] \quad [2, 5] - [1, 3] = [-1, 4]$$

$$[3, 4] \times [2, 2] = [6, 8] \quad [-1, 1] \times [-2, -0.5] = [-2, 2]$$

$$[-1, 1] / [-2, -0.5] = [-2, 2] \quad [4, 10] / [1, 2] = [2, 10]$$

Arithmetic operations on closed intervals satisfy some useful properties. Let us take $A = [a_1, a_2]$, $B = [b_1, b_2]$, $C = [c_1, c_2]$, $D = [d_1, d_2]$, $\mathbf{0} = [0, 0]$, and $\mathbf{1} = [1, 1]$. Then, the properties can be formulated as follows:

1. Commutativity: $A + B = B + A$; $A \times B = B \times A$.
2. Associativity: $(A + B) + C = A + (B + C)$; $(A \times B) \times C = A \times (B \times C)$.
3. Identity: $A = 0 + A = A + 0$; $A = 1 \times A = A \times 1$.
4. Subdistributivity: $A \times (B + C) \subseteq A \times B + A \times C$.
5. Distributivity:
   a. If $b \times c \geq 0$ for every $b \in B$ and $c \in C$, then $A \times (B+C) = A \times B + A \times C$.
   b. If $A = [a, a]$, then $a \times (B + C) = a \times B + a \times C$.
6. $0 \in A - A$ and $1 \in A/A$.
7. Inclusion monotonicity: If $A \subseteq C$ and $B \subseteq D$, then $A * B \subseteq C * D$.

These properties can be readily proven from Eqs. (2.3)–(2.6).

These same equations can be also used to define the arithmetic of fuzzy variables. In fact, as stated, a fuzzy variable can be fully and uniquely represented by its α-cuts, which are indeed closed intervals of real numbers. Therefore, it is possible to apply the interval analysis over each α-cut of the fuzzy variable.

Let A and B denote two fuzzy variables, whose generic α-cuts are A_α and B_α. Let $*$ be any of the four basic arithmetic operations. As A and B are fuzzy variables, $A * B$ is also a fuzzy variable. The α-cuts of the result $A * B$, denoted by $(A * B)_\alpha$, can be easily evaluated from A_α and B_α as

$$(A * B)_\alpha = A_\alpha * B_\alpha$$
(2.7)

for any α between 0 and 1. Let us remember that, when the division is considered, it is required that $0 \notin B_\alpha$ for every α between 0 and 1.

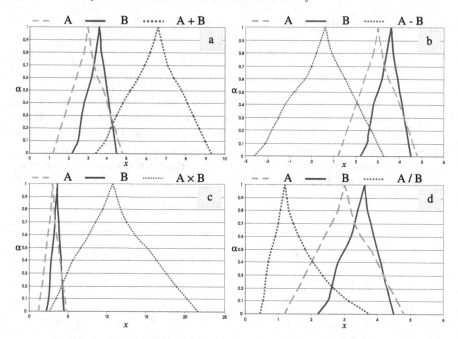

Fig. 2.4. Arithmetic operations on fuzzy variables: (a) addition, (b) subtraction, (c) multiplication, and (d) division.

For the sake of clearness, let us consider the three fuzzy variables A, B, and C, where $C = A * B$, and let us denote the generic α-cuts of A, B, and C as $A_\alpha = [a_1^\alpha, a_2^\alpha]$, $B = [b_1^\alpha, b_2^\alpha]$, and $C = [c_1^\alpha, c_2^\alpha]$, respectively. Then:

- ADDITION: $C = A + B$
  $$[c_1^\alpha, c_2^\alpha] = [a_1^\alpha + b_1^\alpha,\ a_2^\alpha + b_2^\alpha]$$
- SUBTRACTION: $C = A - B$
  $$[\, c_1^\alpha, c_2^\alpha] = [a_1^\alpha - b_2^\alpha,\ a_2^\alpha - b_1^\alpha]$$
- MULTIPLICATION: $C = A \times B$
  $$[c_1^\alpha, c_2^\alpha] = [\min\,(a_1^\alpha b_2^\alpha,\ a_2^\alpha b_1^\alpha,\ a_1^\alpha b_1^\alpha,\ a_2^\alpha b_2^\alpha),\ \max\,(a_1^\alpha b_2^\alpha,\ a_2^\alpha b_1^\alpha,\ a_1^\alpha b_1^\alpha,\ a_2^\alpha b_2^\alpha)]$$
- DIVISION: $C = A/B$
  Provided that $0 \notin [b_1^\alpha, b_2^\alpha]$:
  $$[c_1^\alpha, c_2^\alpha] =$$
  $$[\min\,(a_1^\alpha/b_2^\alpha,\ a_2^\alpha/b_1^\alpha,\ a_1^\alpha/b_1^\alpha,\ a_2^\alpha/b_2^\alpha),\ \max\,(a_1^\alpha/b_2^\alpha,\ a_2^\alpha/b_1^\alpha,\ a_1^\alpha/b_1^\alpha,\ a_2^\alpha/b_2^\alpha)]$$
  for every α between 0 and 1.

Figure 2.4 shows an example of the four arithmetic operations on the two fuzzy variables A and B, represented by the dashed and solid lines, respectively. Starting from the arithmetic operations between fuzzy variables, it is

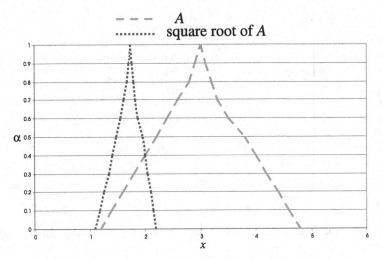

Fig. 2.5. Square root of a fuzzy variable.

also possible to define other mathematical operations. As an example, the square root of a fuzzy variable can be defined as follows.

Let A be a fuzzy variable, defined by its α-cuts $A_\alpha = [a_1^\alpha, a_2^\alpha]$, and let us consider the square root $C = \sqrt{A}$. Provided that the fuzzy number A is positive, that is, $0 \le a_1^\alpha \le a_2^\alpha$ for every α, the generic α-cut of C is

$$[c_1^\alpha, c_2^\alpha] = [\sqrt{a_1^\alpha}, \sqrt{a_2^\alpha}] \tag{2.8}$$

Figure 2.5 shows an example.

It is important to underline that, in some particular applications, it could be also necessary to perform a square root of a fuzzy variable that falls across the zero value. In fact, when the measurement uncertainty of a value near zero is considered, the correspondent confidence interval (and the correspondent fuzzy variable too) contains the zero value. Hence, when the measurement uncertainty of the square root of such a value is considered, the correspondent confidence interval (and the correspondent fuzzy variable too) still contains the zero value. In this case, Eq. (2.8) modifies as follows. If the fuzzy variable is like the one in Fig. 2.6a, that is, the zero value is crossed by the left side of its membership function, it is

$$c_1^\alpha = \begin{cases} -\sqrt{-a_1^\alpha} & \alpha < k^* \\ \sqrt{a_1^\alpha} & \alpha \ge k^* \end{cases}$$

$$c_2^\alpha = \sqrt{a_2^\alpha}$$

where $k^* = \alpha|_{a_1^\alpha = 0}$.

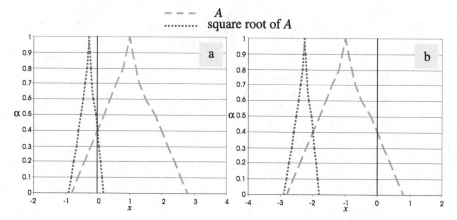

Fig. 2.6. Square root of a fuzzy variable across the zero value. In this example, $k^* = 0.4$.

If the fuzzy variable is like the one in Fig. 2.6b, that is, the zero value is crossed by the right side of its membership function, it is

$$c_1^\alpha = -\sqrt{-a_1^\alpha}$$

$$c_2^\alpha = \begin{cases} -\sqrt{-a_2^\alpha} & \alpha > k^* \\ \sqrt{a_2^\alpha} & \alpha \leq k^* \end{cases}$$

where $k^* = \alpha|_{a_2^\alpha = 0}$.

2.3 A simple example of application of the fuzzy variables to represent measurement results

It has already been stated that a fuzzy variable can be suitably employed to represent a measurement result together with its associated uncertainty [MBFH00, MLF01, UW03, FS03]; in fact, a fuzzy variable can be represented by a set of confidence intervals (the α-cuts) and associated levels of confidence (strictly related to levels α).

Moreover, when an indirect measurement is considered, the mathematics of fuzzy variables allows one to directly obtain the final measurement result in terms of a fuzzy variable. This means that the final measurement result and associated uncertainty can be obtained together, in a single step, which involves fuzzy variables.

In fact, if $y = f(x_1, x_2, \ldots, x_n)$ is the measurement algorithm, the fuzzy variable Y associated with the measurement result and its associated uncertainty is readily given by $Y = f(X_1, X_2, \ldots, X_n)$, where $X_1, X_2, \ldots, X_n$ are the fuzzy variables associated with the input quantities. The operations are of course performed according to the mathematics of fuzzy variables.

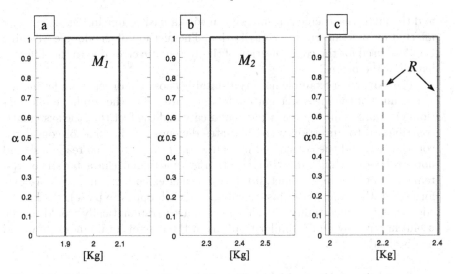

Fig. 2.7. Method of the double weighing. M_1 and M_2 are the two direct measurements, and R is the final result in the two different cases.

Let us consider again the example of the double weighing, and let m_1 and m_2 be 2 kg and 2.4 kg, respectively. The two measurement results can be represented, together with their uncertainty, by two fuzzy variables M_1 and M_2. Let us suppose again that the contributions to uncertainty fall in the interval ±100 g and are not random.

The available information is that the measurement result belongs to the given interval and that each point of the interval is as plausible as the others. Therefore, the fuzzy variables M_1 and M_2 have a rectangular membership function (Fig. 2.7a and b): The confidence interval is given with a level of confidence equal to one, and no other assumptions can be done.

The final measurement result, that is, the arithmetic mean of M_1 and M_2, obtained with the proposed approach, is the fuzzy variable:

$$R = \frac{M_1 + M_2}{2}$$

shown with a solid line in Fig. 2.7c.

Obviously, as stated at the beginning of this chapter, all available information should be used in modeling the measurement process. Therefore, if it was known that the uncertainty contribution behaves systematically and it is due to a difference in length between the two beams of the weighbridge, and thus always takes the (unknown) value a, belonging to interval ±100 g, then the double weighing naturally compensates it, because it affects measurement m_1 with positive sign $(+a)$ and measurement m_2 with negative sign $(-a)$. Hence, the following applies:

$$R = \frac{(m_1 + a) + (m_2 - a)}{2} = \frac{m_1 + m_2}{2}$$

and the final measurement result, shown with a dashed line in Fig. 2.7c, is not affected by this uncertainty contribution. The method of the double weighing is indeed used for the compensation of the systematic error due to the different length of the beams.

This particular example has been suitably chosen in order to underline the importance of the measurement model associated with the available information. Of course, this is a really particular case and the fact that the systematic contribution to uncertainty compensates depends on the kind of considered contribution and the considered measurement procedure. The readers should not erroneously think that compensation is a particular characteristic of systematic contributions! In fact, if the two measurements m_1 and m_2 were performed on the same plate (instead of one on each opposite plate), the knowledge that the uncertainty contribution behaves systematically would not be useful in any way and a similar result as in the case of total ignorance would be obtained.

2.4 Conclusions

In this chapter, fuzzy variables and their mathematics have been defined. Furthermore, it has been shown how they can be used to represent and propagate measurement results each time the uncertainty affecting the results themselves is due to totally unknown contributions, including the systematic ones, when their presence can be only supposed but nothing else can be said about their behavior.

The simple example given in the previous section has shown how immediate is the use of fuzzy variables. When this approach is followed, many advantages can be drawn. Operations are performed directly on fuzzy variables, and no further elaborations are needed in order to process measurement uncertainties.

The measurement result is directly provided in terms of a fuzzy variable, which contains all available information about the measurement result. For instance, if the confidence interval at level of confidence 1 is needed, it simply corresponds to the α-cut at level $\alpha = 0$.

The computational burden is low.

According to their mathematics, the approach based on fuzzy variables can be seen as an attempt to modernize the old theory of errors. In fact, the mathematics are similar (both are based on the interval analysis), but the approach based on fuzzy variables does not refer to the unknowable true value of the measurand.

On the other hand, this approach can be also seen as a complement of the modern Theory of Uncertainty. In fact, it encompasses all concepts of confidence intervals and levels of confidence, while providing a different way to represent incomplete knowledge.

However, the approach based on fuzzy variables must not be considered as an alternative to the statistical approach. In fact, fuzzy variables and random variables are defined in a different way and they obey different mathematics.

Fuzzy variables compose each other according to the mathematics of the intervals; this means, for instance, that the sum of the fuzzy variables A and B (see Fig. 2.4a) is a fuzzy variable whose α-cuts show a width that is the sum of the widths of the corresponding α-cuts of A and B. Therefore, it can be assessed that fuzzy variables are not subject to compensation phenomena. On the other hand, random variables compose each other according to statistics; this means, for instance, that the sum of the random variables A and B is a random variable whose standard deviation is smaller than the sum of the standard deviations of A and B. Hence, random variables are subject to a natural compensation phenomena.

As a result, fuzzy and random variables can be considered to represent different kind of physical phenomena: Fuzzy variables may represent totally unknown phenomena, which physically do not compensate each other, including the systematic ones; random variable may represent, as widely known, random phenomena, which physically compensate each other. Hence, with respect to the Theory of Uncertainty, fuzzy variables and random variables cannot be considered competitive but, rather, complementary. In fact, uncertainty arises in the measurement processes because of the presence of all kinds of uncertainty contributions (random, systematic, unknown). Thus, both approaches are needed.

However, it is not practical and not effective, of course, to follow the two approaches separately. Therefore, a unifying mathematical theory, as well as a unifying mathematical variable, should be found.

This theory is the Theory of Evidence, which encompasses both the probability theory, the mother of random variables, and the possibility theory, the mother of fuzzy variables. The theory of evidence is the topic of Chapter 3.

3

The Theory of Evidence

The mathematical Theory of Evidence was introduced by Shafer in the 1970s as a reinterpretation of Dempster's statistical inference. Shafer's Theory of Evidence begins with the familiar idea of using a number between zero and one to indicate the degree of belief for a proposition on the basis of the available evidence.

For several centuries, the idea of quantifying the degrees of belief by numbers has been identified with the concept of chance. For many mathematicians, the two concepts are uniquely identified by the same concept and name of *probability*. The Theory of Evidence introduced by Shafer rejects this full identification and underlines that numerical degrees of belief and chances have their own roles and obey to different rules that can be derived as particular cases of a more general theoretical framework.

Chances are typical of an *aleatory* (or *random*) *experiment*, like the throw of a die or the toss of a coin. The outcome of such an experiment varies randomly, and the proportion of the times that a particular one of all possible outcomes tends to occur is called the chance (or probability) of that outcome. If X denotes the set of all possible outcomes and, for each possible outcome $x \in X$, the probability $p(x)$ is given (let us suppose set X is finite), then a function $p : X \to [0,1]$, called *probability density*, is specified. By definition, function p satisfies the following conditions:

$$0 \leq p \leq 1$$

$$\sum_{x \in X} p(x) = 1 \tag{3.1}$$

Function $p(x)$ may be zero for some x, meaning that such an outcome is not probable. Of course, there are many possible probability densities on a given set X, and the knowledge of the set X of possible outcomes of an aleatory experiment hardly tells us what probability density governs that experiment.

Sometimes, besides the probability of the different outcomes $x \in X$, one is interested in the proportion of time that the actual outcome falls in a particular subset U of X. This latter proportion is called the chance or probability $\text{Pro}(U)$ of U occurring and is calculated by adding the probabilities of the various elements in U:

$$\text{Pro}(U) = \sum_{x \in U} p(x)$$

Function $\text{Pro}(U)$, called *probability function* (or *chance function*), obviously contains exactly the same information as p. Let now $P(X)$ denote the set of all subsets of X, that is, the power set of X. A function $\text{Pro}(U) : P(X) \to [0,1]$ is a probability function if and only if it obeys the following rules:

$\text{Pro}(\emptyset) = 0$

$\text{Pro}(X) = 1$

if $U, V \subset X$ and $U \cap V = \emptyset$, then $\text{Pro}(U \cup V) = \text{Pro}(U) + \text{Pro}(V)$

These three rules may be called the basic rules for chances. The third one, which says that the probability of disjoint sets add, is called the rule of additivity for chances. It is indeed this rule of additivity that is discarded by Shafer in his Theory of Evidence.

In fact, a *belief function* is a function $\text{Bel} : P(X) \to [0,1]$ that satisfies the following conditions:

$$\text{Bel}(\emptyset) = 0 \qquad (3.2)$$

$$\text{Bel}(X) = 1 \qquad (3.3)$$

and for every positive integer n and every collection $A_1, A_2, \ldots, A_n$ of subsets of X:

$\text{Bel}(A_1 \cup A_2 \cdots \cup A_n)$

$$\geq \sum_i \text{Bel}(A_i) - \sum_{i<j} \text{Bel}(A_i \cap A_j) + \cdots + (-1)^{n+1}\text{Bel}(A_1 \cap A_2 \cdots \cap A_n)$$

$$(3.4)$$

The above relationship can be also written in the following equivalent form:

$$\text{Bel}(A_1 \cup A_2 \cdots \cup A_n) \geq \sum_{\substack{I \subseteq \{1,\ldots,n\} \\ I \neq \emptyset}} (-1)^{|I|+1}\text{Bel}\left(\bigcap_{i \in I} A_i\right)$$

where symbol $|I|$ denotes the *cardinality* of set I, that is the number of elements in I. This means that $(-1)^{|I|}$ is $+1$ if the cardinality of I is even, and -1 if the cardinality is odd.

In the following, Eq. (3.4) will be used to refer indifferently to one of the two preceding equivalent relationships.

If set X is interpreted as a set of possibilities, exactly one of which corresponds to the truth, then belief functions can be used to represent human degrees of belief. In fact, for each subset A of X, the number $\text{Bel}(A)$ can be interpreted as the degree of belief of a person that the truth lies in A and the preceding rules can be considered as rules governing these degrees of belief.

In order to understand the meaning of degree of belief, let us remember Shafer's words as reported in his book [S76]:

'whenever I write of the 'degree of support' that given evidence provides for a proposition or of the 'degree of belief' that an individual accords to the proposition, I picture in my mind an act of judgment. I do not pretend that there exists an objective relation between given evidence and a given proposition that determines a precise numerical degree of support. Nor do I pretend that an actual human being's state of mind with respect to a proposition can ever be described by a precise real number called his degree of belief. Rather, I merely suppose that an individual can make a judgment... he can announce a number that represents the degree to which he judges that evidence to support a given proposition and, hence, the degree of belief he wishes to accord the proposition.'

As revealed by a comparison between the basic rules of probability and the rules for belief functions, it is the rule of additivity that does not apply to degrees of belief. In fact, if $n = 2$, Eq. (3.4) becomes:

$$\text{Bel}(A_1 \cup A_2) \geq \text{Bel}(A_1) + \text{Bel}(A_2) - \text{Bel}(A_1 \cap A_2)$$

and, if $A_1 \cap A_2 = \emptyset$, it becomes:

$$\text{Bel}(A_1 \cup A_2) \geq \text{Bel}(A_1) + \text{Bel}(A_2)$$

which is not as strict as the rule of additivity for chances.

Moreover, if $A_1 = A$ and $A_2 = \bar{A}$ ($\bar{A}$ being the negation of A, that is $A_1 \cap A_2 = 0$ and $A_1 \cup A_2 = X$), the following fundamental property of belief functions is immediately derived:

$$\text{Bel}(A) + \text{Bel}(\bar{A}) \leq 1 \qquad (3.5)$$

In order to fully perceive the meaning of belief functions, let us consider the following simple example, as reported in [S76]. I contemplate a vase that has been represented as a product of the Ming dynasty. Is it genuine or is it counterfeit? Let A_1 correspond to the possibility the vase is genuine and A_2 to the possibility it is counterfeit. Then, $X = \{A_1, A_2\}$ is the set of possibilities, and $P(X) = \{\emptyset, A_1, A_2, X\}$ is its power set. A belief function Bel over X represents my belief if $\text{Bel}(A_1)$ is my degree of belief that the vase is genuine and $\text{Bel}(A_2)$ is my degree of belief that the vase is counterfeit. Let s_1 and s_2 denote, respectively, $\text{Bel}(A_1)$ and $\text{Bel}(A_2)$, for the sake of simplicity. Then,

Eq. (3.5) requires that $s_1 + s_2 \leq 1$. Different pairs of values s_1 and s_2 satisfying this requirement correspond to different situations with respect to the weight of evidence on the two sides of the issue. If the evidence favors the genuineness of the vase, then I will set s_1 near one and s_2 near zero. Substantial evidence on both sides of the issue will lead me to profess some belief on both sides, for example, $s_1 = 0.4$ and $s_2 = 0.3$. If there is little evidence on either side, that is, little reason either to believe or disbelieve the genuineness of the vase, then I will set both s_1 and s_2 very low. In the extreme case of no evidence at all, that is, in the case of total ignorance, I will set both s_1 and s_2 exactly to zero.

This simple example allows us to understand the meaning of the belief function and the way it can be employed. Moreover, it also shows how a belief function can be suitably employed to represent total ignorance. This issue will be considered again in the following discussion.

As stated by Shafer himself, the chances governing an aleatory experiment may or may not coincide with our degree of belief about the outcome of the experiment. If the probabilities are known, then they are surely adopted as our degrees of belief. But if the probabilities are not known, then it will be an exceptional coincidence for our degrees of belief to be equal to them.

However, a proper subclass of the class of belief functions also obeys the rule of additivity and can be used to describe chance in terms of a particular case of numerical degree of belief, so that the differences between the two concepts can be fully perceived, because the same mathematical notation is used. These particular belief functions are called *Bayesian belief functions*. The name derives from the *Bayesian theory*, an extensively developed and very popular theory of partial belief that begins with the explicit premise that all degrees of belief should obey the three rules for chances.

Hence, in the Bayesian theory, belief functions obey to the three basic rules for chances:

$$\mathrm{Bel}(\emptyset) = 0$$

$$\mathrm{Bel}(X) = 1$$

$$\text{if } A_1 \cap A_2 = \emptyset, \text{then } \mathrm{Bel}(A_1 \cup A_2) = \mathrm{Bel}(A_1) + \mathrm{Bel}(A_2)$$

Let us notice that rules 1 and 2 are the same as the first two rules for belief functions, whereas rule 3 is stricter than the third rule for belief functions. This rule is called the Bayes' rule of additivity. It can be readily proven that, if a function obeys to the three preceding rules, it necessarily also obeys the three rules for belief functions, and therefore, it is a belief function. Thus, the functions that obey the three Bayesian rules are a subclass of the belief functions (the *Bayesian belief function*).

The additive degrees of belief of the Bayesian theory correspond to an intuitive picture in which one's total belief is susceptible of division into various portions, and that intuitive picture has two fundamental features: to have a degree of belief in a proposition is to commit a portion of one's belief to it; whenever one commits only a portion of one's belief to a proposition, one must commit the remainder to its negation. However, these features make Bayesian belief functions unsuitable to represent ignorance.

On the contrary, belief functions, which discard the second of these features, can readily represent ignorance, as already seen in the example of the Ming vase. In fact, when we have little evidence about a proposition, the rules of belief functions allow us to express it in a very simple way, by assigning to both that proposition and its negation very low degrees of belief.

Moreover, the rules for belief functions also allow us to assign a zero degree of belief to every subset of a set of possibilities X. This means that our degree of belief is zero for every subset A of the universal set X, and it is one only for X, which contains all possibilities. In other words, the function $\text{Bel} : P(X) \to [0,1]$ defined by

$$\text{Bel}(A) = \begin{cases} 0 & \text{if } A \neq X \\ 1 & \text{if } A = X \end{cases} \tag{3.6}$$

is a belief function: It is known that the correct proposition is contained in X, but it is not known which it is. This belief function is called the *vacuous belief function*, and it is obviously the belief function that represents complete ignorance (or *total ignorance*).

The Bayesian theory, on the other hand, cannot deal so readily with the representation of ignorance, and it is often been criticized on this account. The basic difficulty is that the theory cannot distinguish between lack of belief and disbelief. In other words, it does not allow us to withhold belief from a proposition without according that belief to the negation of the proposition.

Let A and $\bar{A}$ be the proposition and its negation, respectively. Then, the Bayesian theory requires that, because $A \cup \bar{A} = X$, is $\text{Bel}(A \cup \bar{A}) = 1$, and because of the Bayes' rule of additivity, is $\text{Bel}(A) + \text{Bel}(\bar{A}) = 1$. This implies that $\text{Bel}(A)$ cannot be low unless $\text{Bel}(\bar{A})$ is high; that is, failure to believe A needs believing in $\bar{A}$. Therefore, within the Bayesian theory, complete ignorance is expressed as follows:

$$\text{Bel}(A) = \text{Bel}(\bar{A}) = \frac{1}{2}$$

that is, with an equal degree of belief for both A and $\bar{A}$.

Although this solution seems to be plausible in case only two possibilities are considered, that is $(A, \bar{A})$, it is totally useless in representing ignorance when the set of possibilities contains more than two elements, as demonstrated by the following example.

Are there or are there not living beings in orbit around the star Sirius? Some scientists may have evidence on this question, but most of us will profess complete ignorance about it. So, if A_1 corresponds to the possibility that there is life and A_2 to the possibility that there is not, we will adopt, according to Shafer's Theory of Evidence, the vacuous belief function over the set of possibilities $X = \{A_1, A_2\}$; that is, $\mathrm{Bel}(A_1) = \mathrm{Bel}(A_2) = 0$ and $\mathrm{Bel}(X) = 1$. On the other side, the Bayesian theory will set a degree of belief $\frac{1}{2}$ to both A_1 and A_2.

But this issue may be also reconsidered, by adding, for example, the question of whether planets around Sirius even exist. Then, we have the set of possibilities $\Omega = \{B_1, B_2, B_3\}$ where B_1 corresponds to the possibility that there is life around Sirius, B_2 corresponds to the possibility that there are planets but not life, and B_3 corresponds to the possibility that there are not even planets. Of course sets X and Ω are related to each other, because A_1 corresponds to B_1 and A_2 corresponds to $\{B_2, B_3\}$. The believers in the Theory of Evidence who profess complete ignorance about that issue will adopt a vacuous belief function over Ω, and this is consistent with having adopted a vacuous belief function over X. But the Bayesian will find it difficult to specify consistent degrees of belief, over both X and Ω, representing ignorance. In fact, focusing on X, he might claim that ignorance is represented by

$$\mathrm{Bel}(A_1) = \mathrm{Bel}(A_2) = \frac{1}{2}$$

whereas focusing on Ω, he might claim that the best he can do to represent ignorance is to set

$$\mathrm{Bel}(B_1) = \mathrm{Bel}(B_2) = \mathrm{Bel}(B_3) = \frac{1}{3}$$

But this yields to

$$\mathrm{Bel}(B_1) = \frac{1}{3} \quad \text{and} \quad \mathrm{Bel}(B_2, B_3) = \frac{2}{3}$$

which are inconsistent with the assessment done when considering X.

Therefore, the Bayesian theory is inadequate for the representation of total ignorance. The same conclusion can also be drawn for the probability theory, because the basic rules for chances are exactly the same as the three Bayesian rules for belief functions.

Thus, Shafer's Theory of Evidence is the only mathematical theory able to represent total ignorance. As stated in Chapter 2, total ignorance is a frequent situation in the field of the measurement practice. In fact, every time the only available information is that the measurement result falls within a given interval (a confidence interval), without any additional information about its distribution, we are exactly in the case of total ignorance. In Shafer's Theory of Evidence, propositions can always be represented as subsets of a given set; hence, the given interval is the set of possibilities X, and the different possibilities are the subintervals of X. Therefore, as there is no evidence about the belonging to any of the subintervals of X, this means that the degree of belief for each subinterval of X is zero, and therefore, the vacuous belief function over X must be adopted.

The representation of propositions as well as sets is very easily adopted in the case of measurement results. Let x denote the measurement result and X be the set of its possible values: We are obviously interested in propositions of the form 'the measurement result x is in T' where T is a subset of X. Thus, the propositions of our interest are in a one-to-one correspondence with the subsets of X, and the set of all propositions of interest corresponds to the set of all subsets of X, which is denoted $P(X)$ [S76]. X is called the *frame of discernment*.

One reason why the correspondence between propositions and subsets is useful is that it translates the logical notions of conjunction, disjunction, implication, and negation into the more graphic notions of intersection, union, inclusion, and complementation. Indeed, if A and B are two subsets of X, and A' and B' are the corresponding propositions, then the intersection $A \cap B$ corresponds to the conjunction of A' and B'; the union $A \cup B$ corresponds to the disjunction of A' and B'; A is a subset of B if and only if A' implies B'; and A is the complement of B with respect to X if and only if A' is the negation of B'.

3.1 Basic definitions

In the introduction to this chapter, it has been briefly explained the difference among belief functions, probability functions, and Bayesian belief functions, in order to prevent possible confusion between concepts that are different, although similar.

In this section, we focus on belief functions and give a mathematical form to the intuitive concept in which a portion of belief can be committed to a proposition but does not need to be committed either to it or to its negation.

In a strict logical approach, the belief function expresses the concept that the portion of belief committed to a proposition is also committed to any other proposition it implies. In more mathematical terms, if the frame of discernment is considered, this means that a portion of belief committed to one subset is also committed to any of its subsets.

Under this approach, it could be useful, whenever this is possible, to partition the belief assigned to a set X among the different subsets A of X, so that to each subset A, it is assigned that portion of belief that is committed exactly to A and to nothing smaller. This can be done if a function m, called *basic probability assignment*, is defined.

Let X be the frame of discernment: The basic probability assignment is then defined as a function $m : P(X) \rightarrow [0,1]$ that satisfies the following conditions:

$$m(\emptyset) = 0 \tag{3.7}$$

$$\sum_{A \in P(X)} m(A) = 1 \tag{3.8}$$

Condition (3.7) reflects the fact that no belief must be committed to $\emptyset$, whereas Eq. (3.8) reflects the convention that the total belief has unitary value.

For each set $A \in P(X)$, quantity $m(A)$ is called the *basic probability number* and represents the measure of the belief that is committed exactly to A; that is, $m(A)$ refers only to set A and does not imply anything about the degrees of belief committed to the subsets of A.

When measurement results are considered, the basic probability number $m(A)$ can be interpreted as the degree of belief that the measurement result falls within interval A; but $m(A)$ does not provide any further evidence in support to the belief that the measurement result belongs to any of the various subintervals of A. This means that, if there is some additional evidence supporting the claim that the measurement result falls within a subinterval of A, say $B \subset A$, it must be expressed by another value $m(B)$.

Although Eq. (3.8) looks similar to Eq. (3.1) for probability density functions, there is a fundamental difference between probability density functions and basic probability assignment functions: The former are defined on X, and the latter are defined on $P(X)$. This means that the probability density function is defined on every singleton of X, whereas the basic probability assignment function can be defined on the various subsets of X.

Hence, it can be stated that the basic probability assignment function is a more general function than the probability density function, which is only defined on very particular subsets of X: the singletons.

From Eqs. (3.7) and (3.8), the following characteristics of basic probability assignments can be drawn:

- it is not required that $m(X) = 1$;
- it is not required that $m(A) \leq m(B)$ when $A \subseteq B$;
- there is no relationship between $m(A)$ and $m(\overline{A})$.

As quantity $m(A)$ measures the belief committed exactly to A, in order to obtain the measure of the total belief committed to A, that is the belief committed to A and all its subsets, the basic probability numbers associated with all subsets B of A must be added to:

$$t(A) = \sum_{B|\ B \subseteq A} m(B) \tag{3.9}$$

It can be proved that the total belief $t(A)$ committed to A is a belief function; that is, it satisfies Eqs. (3.2) to (3.4).

Proof:

1. If $A = \emptyset$, then by taking into account Eq. (3.7):

$$t(\emptyset) = \sum_{B|\ B \subseteq \emptyset} m(B) = m(\emptyset) = 0$$

and Eq. (3.2) is satisfied.

2. If $A = X$, then by taking into account Eq. (3.8):

$$t(X) = \sum_{B| \ B \subseteq X} m(B) = \sum_{B \in P(X)} m(B) = 1$$

and Eq. (3.3) is satisfied.

3. Let us consider a fixed collection $A_1, A_2, \ldots, A_n$ of subsets of X, and let us also set, for each $B \subset X$, $I(B) = \{ i | 1 \leq i \leq n; B \subset A_i \}$. This last relationship shows that $I(B) \neq \emptyset$ only if B is a subset of at least one of the sets $A_1, A_2, \ldots, A_n$, as also shown in Fig. 3.1. Let us consider the following term:

$$\sum_{\substack{I \subseteq \{1,\ldots,n\} \\ I \neq \emptyset}} (-1)^{|I|+1} t \left(\bigcap_{i \in I} A_i \right)$$

From Eq. (3.9), it can be written as follows:

$$\sum_{\substack{I \subseteq \{1,\ldots,n\} \\ I \neq \emptyset}} (-1)^{|I|+1} t \left(\bigcap_{i \in I} A_i \right) = \sum_{\substack{I \subseteq \{1,\ldots,n\} \\ I \neq \emptyset}} (-1)^{|I|+1} \sum_{B \subseteq \bigcap_{i \in I} A_i} m(B)$$

The two sums at the second member of the above relationship can be reversed in order as follows:

$$\sum_{\substack{I \subseteq \{1,\ldots,n\} \\ I \neq \emptyset}} (-1)^{|I|+1} t \left(\bigcap_{i \in I} A_i \right) = \sum_{\substack{B \subseteq X \\ I(B) \neq \emptyset}} m(B) \sum_{\substack{I \subseteq I(B) \\ I \neq \emptyset}} (-1)^{|I|+1}$$

In fact, as also shown in Fig. 3.1, considering all possible sets B, included in the intersection of one or more sets A_i, is equivalent to consider all possible subsets B of X, for which $I(B) \neq \emptyset$. Moreover, the proper choice of sets B forces set I, which now appears in the internal sum, to be included in $I(B)$.

Let us now consider this internal sum:

$$\sum_{\substack{I \subseteq I(B) \\ I \neq \emptyset}} (-1)^{|I|+1} = \sum_{\substack{I \subseteq I(B) \\ I \neq \emptyset}} (-1)^{|I|}(-1)^{+1} = - \sum_{\substack{I \subseteq I(B) \\ I \neq \emptyset}} (-1)^{|I|}$$

$$= - \left[\sum_{I \subseteq I(B)} (-1)^{|I|} - (+1) \right] = 1 - \sum_{I \subseteq I(B)} (-1)^{|I|}$$

Hence,

$$\sum_{\substack{I \subseteq \{1,\ldots,n\} \\ I \neq \emptyset}} (-1)^{|I|+1} t \left(\bigcap_{i \in I} A_i \right) = \sum_{\substack{B \subseteq X \\ I(B) \neq \emptyset}} m(B) \left(1 - \sum_{I \subseteq I(B)} (-1)^{|I|} \right)$$

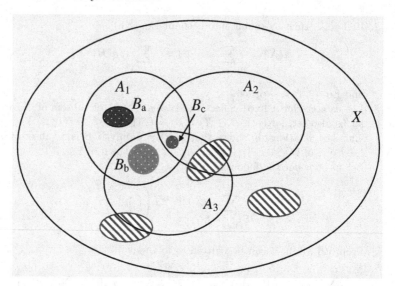

Fig. 3.1. A_1, A_2, and A_3 are subsets of the universal set X. Sets B_a, B_b, and B_c are included at least in one A_i. Hence, for them, it is $I(B) \neq \emptyset$. Other sets, like the ones with dashed background, are not included in any A_i. For them, it is $I(B) = \emptyset$.

Let us now consider the term $\sum_{I \subseteq I(B)}(-1)^{|I|}$. It can be proved that

$$\sum_{B \subseteq A}(-1)^{|B|} = \begin{cases} 1 & \text{if } A = \emptyset \\ 0 & \text{otherwise} \end{cases}$$

In fact, let us first consider the case $A = \emptyset$; that is, $|A| = 0$. In this case,

$$\sum_{B \subseteq A}(-1)^{|B|} = (-1)^{|A|} = (-1)^0 = 1$$

Let us now consider the general situation where $|A| = n > 0$. Hence, B may have cardinalities $0, 1, 2, \ldots, n$. Moreover, only one subset B exists with cardinality 0, which contributes with a value $(-1)^0$; n sets B exist with cardinality 1, which contribute with a value $(-1)^1$ each; $\binom{n}{2}$ sets B exist with cardinality 2, which contribute with a value $(-1)^2$ each; and $\binom{n}{n} = 1$ set B exists with cardinality n, which contributes with a value $(-1)^n$. Hence,

$$\sum_{B \subseteq A}(-1)^{|B|} = \sum_{k=0}^{n}(-1)^k \binom{n}{k} = \sum_{k=0}^{n}(-1)^k \binom{n}{k}(+1)^{n-k}$$

Let us now remember the binomial theorem:

$$(a+b)^n = \sum_{k=0}^{n} a^k b^{n-k} \binom{n}{k}$$

Then, it can be recognized that the second member of the above relationship is equal to the binomial theorem with $a = -1$ and $b = +1$. Hence,

$$\sum_{B \subseteq A} (-1)^{|B|} = (-1+1)^n = 0 \quad \forall n > 0$$

Therefore, being $I(B) \neq \emptyset$, as forced by the external sum, it is $\sum_{I \subseteq I(B)} (-1)^{|I|} = 0$ and

$$\sum_{\substack{I \subseteq \{1,\ldots,n\} \\ I \neq \emptyset}} (-1)^{|I|+1} \, t \left(\bigcap_{i \in I} A_i \right) = \sum_{\substack{B \subseteq X \\ I(B) \neq \emptyset}} m(B)$$

As shown in Fig. 3.1, it is surely:

$$\sum_{\substack{B \subseteq X \\ I(B) \neq \emptyset}} m(B) \leq \sum_{B \subseteq (A_1 \cup A_2 \cdots \cup A_n)} m(B) = t(A_1 \cup A_2 \cdots \cup A_n)$$

Hence, function t satisfies to Eq. (3.4) too. It can be concluded that function t is a belief function.

Hence, given a basic probability assignment $m : P(X) \to [0,1]$, a belief function $\text{Bel} : P(X) \to [0,1]$ is uniquely determined by

$$\text{Bel}(A) = \sum_{B \mid B \subseteq A} m(B) \tag{3.10}$$

The belief function with the simplest structure is surely the vacuous belief function (3.6), which expresses no evidence (or total ignorance). This belief function is obtained by the basic probability assignment:

$$m(A) = \begin{cases} 0 & \text{if } A \neq X \\ 1 & \text{if } A = X \end{cases}$$

Equivalently, given a belief function $\text{Bel} : P(X) \to [0,1]$, the basic probability assignment $m : P(X) \to [0,1]$ is uniquely determined by

$$m(A) = \sum_{B \mid B \subseteq A} (-1)^{|A-B|} \, \text{Bel}(B) \tag{3.11}$$

Equation (3.11) can be directly derived from Eq. (3.10) and vice versa.

Proof:

- Let us suppose that Eq. (3.10) applies for all $A \in X$. Then, let us consider the term:

$$\sum_{B \subseteq A} (-1)^{|A-B|} \text{Bel}(B)$$

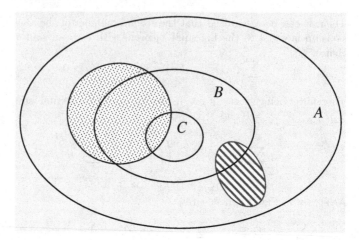

Fig. 3.2. The three sets A, B, and C show the identity of the two following propositions. For every set B, included in A, let us consider all sets C included in B. Sets like the one with a dashed background are never considered. For every set C, included in A, let us consider all sets B, which satisfy to $C \subseteq B \subseteq A$. Sets like the one with a dotted background are never considered.

It can be readily proven that, if $B \subset A$, as required by the sum in the above term, then $|A - B| = |A| - |B|$ and $(-1)^{|A-B|} = (-1)^{|A|} (-1)^{|B|}$. Therefore, considering also Eq. (3.10):

$$\sum_{B \subseteq A} (-1)^{|A-B|} \mathrm{Bel}(B) = (-1)^{|A|} \sum_{B \subseteq A} (-1)^{|B|} \mathrm{Bel}(B)$$

$$= (-1)^{|A|} \sum_{B \subseteq A} (-1)^{|B|} \sum_{C \subseteq B} m(C)$$

The two sums at the second member of the above relationship can be reversed in order as follows:

$$\sum_{B \subseteq A} (-1)^{|A-B|} \mathrm{Bel}(B) = (-1)^{|A|} \sum_{C \subseteq A} m(C) \sum_{\substack{B \\ C \subseteq B \subseteq A}} (-1)^{|B|}$$

In fact, as also shown in Fig. 3.2, considering all sets C included in B, for every set B included in A, is equivalent to consider all sets B, which satisfy to $C \subseteq B \subseteq A$, for every set C included in A. Let us now consider the internal sum: $\sum_{\substack{B \\ C \subseteq B \subseteq A}} (-1)^{|B|}$. It can be proved that

$$\sum_{\substack{B \\ C \subseteq B \subseteq A}} (-1)^{|B|} = \begin{cases} (-1)^{|A|} & \text{if } A = C \\ 0 & \text{otherwise} \end{cases}$$

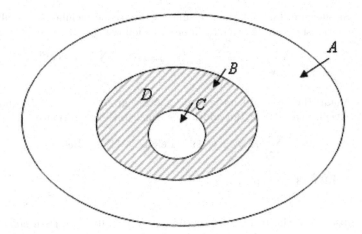

Fig. 3.3. The three oval sets A, B, and C satisfy to $C \subseteq B \subseteq A$. Set D is the set with dashed background. It is $B = C \cup D$.

In this respect, let us now consider Fig. 3.3. As $C \subseteq B$, then every set B can be seen as $B = C \cup D$, where D is a set included in $A - C$. Hence,

$$\sum_{\substack{B \\ C \subseteq B \subseteq A}} (-1)^{|B|} = \sum_{D \subseteq (A-C)} (-1)^{|C \cup D|} = (-1)^{|C|} \sum_{D \subseteq (A-C)} (-1)^{|D|}$$

which proves the above equation because, as already proved, it is

$$\sum_{D \subseteq (A-C)} (-1)^{|D|} = \begin{cases} 1 & \text{if } A = C \\ 0 & \text{otherwise} \end{cases}$$

This means that, in the main equation, the only term in the external sum different from zero is that where $A = C$, for which the internal sum takes value $(-1)^{|A|}$. This leads to

$$\sum_{B \subseteq A} (-1)^{|A-B|} \text{Bel}(B) = (-1)^{|A|} \, m(A) \, (-1)^{|A|} = m(A)$$

and Eq. (3.11) is proved.

Let us suppose that Eq. (3.11) applies for all $A \in X$. Then, let us consider the term:

$$\sum_{B \subseteq A} m(B)$$

By applying Eq. (3.11):

$$\sum_{B \subseteq A} m(B) = \sum_{B \subseteq A} \sum_{C \subseteq B} (-1)^{|B-C|} \text{Bel}(C)$$

As shown in Fig. 3.2, the two sums at the second member of the above relationship can be reversed in order as follows:

$$\sum_{B \subseteq A} m(B) = \sum_{C \subseteq A} (-1)^{|C|} \text{Bel}(C) \sum_{\substack{B \\ C \subseteq B \subseteq A}} (-1)^{|B|}$$

Again, the internal sum is zero for every group of sets A, B, and C, except when $A = C$ (for which it is equal to $(-1)^{|A|}$). Hence,

$$\sum_{B \subseteq A} m(B) = (-1)^{|A|} \text{Bel}(A) (-1)^{|A|} = \text{Bel}(A)$$

and Eq. (3.10) is proved.

A subset A of the frame of discernment X is called the *focal element* of a belief function Bel over X if $m(A) \geq 0$. As this name suggests, focal elements are subsets of X on which the available evidence focuses. The union of all focal elements of a belief function is called its *core*. The pair $\{\Im, \mu\}$, where $\Im$ and μ denote the focal elements and the associated basic probability assignment, respectively, is often called a *body of evidence*. Therefore, if the frame X is finite, it can be fully characterized by a list of its bodies of evidence.

Until now, degrees of beliefs have been fully described. However, the belief of one person about a proposition A is not fully described by his degree of belief $\text{Bel}(A)$, because $\text{Bel}(A)$ does not reveal to what extent the person doubts A, or, in other words, to what extent he believes its negation $\bar{A}$. A full description about the belief of a person consists of the degree of belief $\text{Bel}(A)$ together with the *degree of doubt*:

$$\text{Dou}(A) = \text{Bel}(\bar{A})$$

If $\text{Bel}(A)$ expresses the degree of belief about the fact that event A occurs, the degree of doubt $\text{Dou}(A)$ expresses the degree of belief that event A does not occur.

The degree of doubt is seldom used, being less useful than the quantity:

$$\text{Pl}(A) = 1 - \text{Dou}(A)$$

which expresses the extent to which one fails to doubt A, in other words, the extent to which one finds A plausible. Thus, $\text{Pl}(A)$ is called *plausibility* or *upper probability* of A.

Whenever Bel is a belief function over the frame of discernment X, the function $\text{Pl} : P(X) \to [0, 1]$ defined by

$$\text{Pl}(A) = 1 - \text{Bel}(\bar{A}) \tag{3.12}$$

is called the *plausibility function* (or *upper probability function*) for Bel.

As, for all $A \in P(X)$, Eq. (3.12) can be rewritten as

$$\text{Bel}(A) = 1 - \text{Pl}(\bar{A})$$

then, it can be concluded that functions Bel and Pl contain precisely the same information.

From Eqs. (3.8), (3.10), and (3.12), it is possible to obtain the plausibility function expressed in terms of the basic probability assignment:

$$\text{Pl}(A) = \sum_{B|\ B\cap A \neq \emptyset} m(B) \tag{3.13}$$

Proof:

Considering Eq. (3.8) and (3.12):

$$\text{Pl}(A) = 1 - \text{Bel}(\bar{A}) = 1 - \sum_{B \subseteq \bar{A}} m(B)$$

$$= \sum_{B \in P(X)} m(B) - \sum_{B \subseteq \bar{A}} m(B) = \sum_{B|\ B\cap A \neq \emptyset} m(B)$$

In fact, if at the sum of $m(B)$ for every B belonging to the power set, we have to subtract those for which B are included in $\bar{A}$, it means that the sum of $m(B)$ for only sets B that overlap with A must be considered.

By comparing (3.10) and (3.13), it follows, for every $A \in P(X)$:

$$\text{Pl}(A) \geq \text{Bel}(A)$$

Proof:

The above equation can be simply demonstrated by considering that every $B|B \subseteq A$ also satisfy to $B \cap A \neq \emptyset$, whereas the opposite does not apply.

The above definitions show that, given a frame X, belief functions, plausibility functions, and basic probability assignments are all dependent on each other. Hence, from the knowledge of only one among the three functions, it is possible to deduce the other two.

According to this statement, total ignorance can be also represented by the vacuous plausibility function; Eq. (3.12) leads to

$$\text{Pl}(A) = \begin{cases} 0 & \text{if } A = \emptyset \\ 1 & \text{if } A \neq \emptyset \end{cases}$$

As with the belief functions, plausibility functions can also be defined in terms of a set of conditions to which they must satisfy. A function $\text{Pl} : P(X) \to [0,1]$ is a plausibility function if it satisfies the following conditions:

$$Pl(\emptyset) = 0 \tag{3.14}$$

$$Pl(X) = 1 \tag{3.15}$$

and

$$Pl(A_1 \cap A_2 \cdots \cap A_n)$$
$$\geq \sum_j Pl(A_j) - \sum_{j<k} Pl(A_j \cup A_k) + \cdots + (-1)^{n+1}Pl(A_1 \cup A_2 \cdots \cup A_n) \tag{3.16}$$

The above relationship can be also written in the following equivalent form:

$$Pl(A_1 \cap A_2 \cdots \cap A_n) \geq \sum_{\substack{I \subseteq \{1,\ldots,n\} \\ I \neq \emptyset}} (-1)^{|I|+1} Pl\left(\bigcup_{i \in I} A_i\right)$$

In the following, Eq. (3.16) will be used to refer indifferently to one of the two preceding equivalent relationships.

It can be demonstrated that the function defined by Eq. (3.13) satisfies Eqs. (3.14) to (3.16).

Proof:

If (3.13) applies:
1. If $A = \emptyset$, then

$$\sum_{B\mid \emptyset \cap B \neq \emptyset} m(B) = 0$$

because the intersection with the empty set always gives an empty set. Hence, Eq. (3.14) is satisfied.
2. If $A = X$, then

$$\sum_{B\mid B \cap X \neq \emptyset} m(B) = \sum_{B \subseteq P(X)} m(B) = 1$$

and Eq. (3.15) is satisfied.
3. Let us consider a fixed collection $A_1, A_2, \ldots, A_n$ of subsets of X. Let us also set, for each $B \subseteq X$,

$$I'(B) = \{i \mid 1 \leq i \leq n, B \cap A_i \neq \emptyset\}$$

that is, $I'(B) \neq \emptyset$ only if B overlaps with at least one of the sets $A_1, A_2, \ldots, A_n$, as also shown in Fig. 3.4.

Let us now consider the second term of Eq. (3.16). By taking into account Eq. (3.13):

$$\sum_{\substack{I \subseteq \{1,\ldots,n\} \\ I \neq \emptyset}} (-1)^{|I|+1} Pl\left(\bigcup_{i \in I} A_i\right) = \sum_{\substack{I \subseteq \{1,\ldots,n\} \\ I \neq \emptyset}} (-1)^{|I|+1} \sum_{B\mid (B \cap \bigcup_{i \in I} A_i) \neq \emptyset} m(B)$$

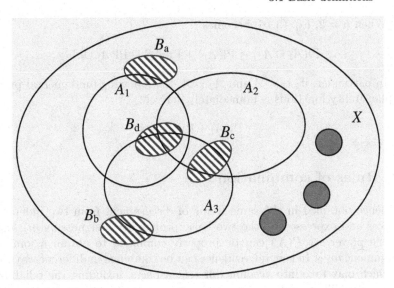

Fig. 3.4. A_1, A_2, and A_3 are subsets of the universal set X. Sets B_a, B_b, B_c, and B_d are examples of sets that overlap with at least one A_i. Hence, for them it is $I'(B) \neq \emptyset$. Only sets completely outside each A_i (like the gray ones) have $I'(B) = \emptyset$. However, only set B_d overlaps with $A_1 \cap A_2 \cap A_3$.

Let us now consider that, if B overlaps with $\bigcup_{i \in I} A_i$, as required by the internal sum, then B overlaps with some of the considered sets A_i, as also shown in Fig. 3.4. Under this assumption, $I'(B) \subseteq I$ and $I'(B) \neq \emptyset$.

Hence, it is possible to reverse in order the two sums as follows:

$$\sum_{\substack{I \subseteq \{1,\dots,n\} \\ I \neq \emptyset}} (-1)^{|I|+1} \, \mathrm{Pl}\left(\bigcup_{i \in I} A_i\right) = \sum_{\substack{B \subseteq X \\ I'(B) \neq \emptyset}} m(B) \sum_{\substack{I \subseteq I'(B) \\ I \neq \emptyset}} (-1)^{|I|+1}$$

As proved, whenever $I'(B) \neq \emptyset$, the internal sum is equal to one. Hence,

$$\sum_{\substack{I \subseteq \{1,\dots,n\} \\ I \neq \emptyset}} (-1)^{|I|+1} \, \mathrm{Pl}\left(\bigcup_{i \in I} A_i\right) = \sum_{\substack{B \subseteq X \\ I'(B) \neq \emptyset}} m(B)$$

As shown in Fig. 3.4, it is surely

$$\sum_{\substack{B \subseteq X \\ I'(B) \neq \emptyset}} m(B) \geq \sum_{\substack{(B \cap \bigcap\limits_{i \in \{1,\dots,n\}} A_i) \neq \emptyset}} m(B) = \mathrm{Pl}(A_1 \cap A_2 \cdots \cap A_n)$$

Therefore, Eq. (3.16) is also satisfied.

When $n = 2$, Eq. (3.16) becomes

$$Pl(A_1 \cap A_2) \geq Pl(A_1) + Pl(A_2) - Pl(A_1 \cup A_2)$$

In particular, if $A_1 = A$ and $A_2 = \bar{A}$, the following fundamental property of plausibility functions is immediately derived:

$$Pl(A) + Pl(\bar{A}) \geq 1 \tag{3.17}$$

3.2 Rules of combination

Evidence obtained in the same frame of discernment from two independent sources and expressed by the two basic probability assignments m_1 and m_2 on the power set $P(X)$ can be properly combined to obtain a joint basic assignment $m_{1,2}$. In general, evidence can be combined in different ways, some of which may take into account different aspects, including the reliability of the sources. The standard way to combine evidence is the *Dempster's rule of combination*, expressed by [KY95], [S76]:

$$m_{1,2} = \begin{cases} \dfrac{\displaystyle\sum_{B \cap C = A} m_1(B)\, m_2(C)}{1 - K} & \text{if } A \neq \emptyset \\ 0 & \text{if } A = \emptyset \end{cases} \tag{3.18}$$

where

$$K = \sum_{B \cap C = \emptyset} m_1(B)\, m_2(C) \tag{3.19}$$

According to Dempster's rule of combination, the degree of evidence $m_1(B)$ from the first source, which focuses on set $B \in P(X)$, and the degree of evidence $m_2(C)$ from the second source, which focuses on set $C \in P(X)$, are combined by considering the product $m_1(B)\, m_2(C)$, which focuses on the intersection $B \cap C$. Obviously, as there could be more than one pair of sets that have the same intersection, a sum of products $m_1(B)\, m_2(C)$ is present in Eq. (3.18). Moreover, some intersections may be empty. Therefore, in order to satisfy condition (3.7) for the basic probability assignments, products $m_1(B)\, m_2(C)$ for all focal elements B of m_1 and all focal elements C of m_2 such that $B \cap C = \emptyset$, expressed by Eq. (3.19), must not be included in the definition of the joint basic assignment $m_{1,2}$. But this means that the sum of products for all focal elements B of m_1 and all focal elements C of m_2 such that $B \cap C \neq \emptyset$ is not equal to one, but to $1 - K$. This value is therefore present at the denominator of Eq. (3.18), in order to satisfy the normalization condition (3.8) for the basic probability assignments.

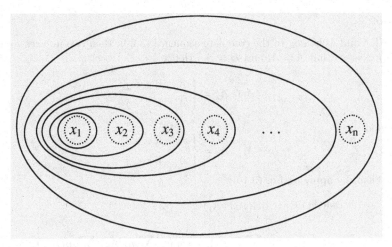

Fig. 3.5. Complete sequence of nested subsets of X.

3.3 Possibility theory

Most of the basic definitions given in the previous sections depend on the focal elements. Therefore, adding suitable constraints to the focal elements may lead to some interesting particular cases of the Theory of Evidence.

A first, interesting case is given by the possibility theory, which deals with bodies of evidence whose focal elements are nested; that is, focal elements can be ordered so that each one is contained in the following one.

Nested focal elements are also called, for the sake of brevity, *consonants*. Similarly, a belief function whose focal elements are nested is said to be *consonant*.

Let $X = \{x_1, x_2, \ldots, x_n\}$ be the frame of discernment or the universal set. The complete sequence of nested subsets of X is given by

$$A_i = \{x_1, x_2, \ldots, x_i\} \tag{3.20}$$

for $i = 1, \ldots, n$.

In fact, sets A_i defined by Eq. (3.20) satisfy the following relationship:

$$A_1 \subset A_2 \subset \cdots \subset A_n \equiv X \tag{3.21}$$

as shown in Fig. 3.5.

Hence, within the possibility theory, the focal elements of X are some or all of the subsets in the complete sequence in Eq. (3.20).

In this particular situation, belief and plausibility functions satisfy the following relationships, for all focal elements A and $B \in P(X)$:

$$\mathrm{Bel}(A \cap B) = \min[\mathrm{Bel}(A), \mathrm{Bel}(B)] \tag{3.22}$$

$$\mathrm{Pl}(A \cup B) = \max[\mathrm{Pl}(A), \mathrm{Pl}(B)] \tag{3.23}$$

Proof:

If A and B belong to the complete sequence (3.20), then two indexes i and j exist so that $A = A_i$ and $B = A_j$. Hence,

$$A_i \cap A_j = \begin{cases} A_i & \text{if } i < j \\ A_j & \text{if } i > j \end{cases}$$

and

$$A_i \cup A_j = \begin{cases} A_i & \text{if } i > j \\ A_j & \text{if } i < j \end{cases}$$

Then, by applying Eq. (3.10):

$$\text{Bel}(A \cap B) = \text{Bel}(A_i \cap A_j)$$

$$= \sum_{k=1}^{\min(i,j)} m(A_k) = \min\left[\sum_{k=1}^{i} m(A_k), \sum_{k=1}^{j} m(A_k)\right]$$

$$= \min\left[\text{Bel}(A_i), \text{Bel}(A_j)\right] = \min\left[\text{Bel}(A), \text{Bel}(B)\right]$$

and by applying Eq. (3.12):

$$\text{Pl}(A \cup B) = 1 - \text{Bel}(\overline{A \cup B}) = 1 - \text{Bel}(\overline{A} \cap \overline{B})$$

$$= 1 - \min\left[\text{Bel}(\overline{A}), \text{Bel}(\overline{B})\right] = \max\left[1 - \text{Bel}(\overline{A}), 1 - \text{Bel}(\overline{B})\right]$$

$$= \max\left[\text{Pl}(A), \text{Pl}(B)\right]$$

3.3.1 Necessity and possibility functions

When a belief function satisfies Eq. (3.22) and the corresponding plausibility function satisfies Eq. (3.23), they are called *necessity function* (Nec) and *possibility function* (Pos), respectively. Hence, Eqs. (3.22) and (3.23) become

$$\text{Nec}(A \cap B) = \min[\text{Nec}(A), \text{Nec}(B)] \tag{3.24}$$

$$\text{Pos}(A \cup B) = \max[\text{Pos}(A), \text{Pos}(B)] \tag{3.25}$$

for all A and $B \in P(X)$.

Equations (3.24) and (3.25) can be also used, in the possibility theory, to define the necessity and possibility functions axiomatically; then, the nested structure of their focal elements can be proven to be a necessary and sufficient condition by a theorem in this theory. This approach is only briefly recalled here for the sake of completeness, because it allows us to define the necessity and possibility functions also for not finite universal sets.[1] This general formulation is based on the following two definitions, which extend Eqs. (3.24) and (3.25) to the case of not finite universal sets.

---

[1] A not finite universal set does not mean that the set is infinite, but that it contains an infinite number of elements.

- Let us consider the universal set X and its power set $P(X)$. Let K be an arbitrary index set. It can be proved [KY95] that a function Nec is a necessity function if and only if it satisfies the following relationship:

$$\text{Nec}\left(\bigcap_{k \in K} A_k \right) = \inf_{k \in K} \text{Nec}(A_k) \tag{3.26}$$

for any family $\{A_k | \ k \in K\}$ in $P(X)$.

- Let us consider the universal set X and its power set $P(X)$. Let K be an arbitrary index set. It can be proved [KY95] that a function Pos is a possibility function if and only if it satisfies the following relationship:

$$\text{Pos}\left(\bigcup_{k \in K} A_k \right) = \sup_{k \in K} \text{Nec}(A_k) \tag{3.27}$$

for any family $\{A_k | \ k \in K\}$ in $P(X)$.

Having stated that the necessity and possibility functions can be defined in the more general case of sets with an infinite number of elements, let us now go back to the simpler case of finite sets, where all mathematical definitions are much simpler. This can be done without lacking generality, because all mathematical definitions can be readily extended to the infinite sets thanks to Eqs. (3.26) and (3.27).

As necessity functions are special belief functions and possibility functions are special plausibility functions, they must satisfy Eqs. (3.5), (3.12), and (3.17). Hence,

$$\text{Nec}(A) + \text{Nec}(\bar{A}) \leq 1 \tag{3.28}$$

$$\text{Pos}(A) + \text{Pos}(\bar{A}) \geq 1 \tag{3.29}$$

$$\text{Nec}(A) = 1 - \text{Pos}(\bar{A}) \tag{3.30}$$

Considering Eqs. (3.10) and (3.13), and the sequence of nested subsets in Eq. (3.20), for which $A_1 \subset A_2 \subset \cdots \subset A_n \equiv X$, necessity and possibility functions can be expressed, in terms of the basic probability assignment, as

$$\text{Nec}(A_j) = \sum_{A_k | A_k \subseteq A_j} m(A_k) = \sum_{k=1}^{j} m(A_k) \tag{3.31}$$

$$\text{Pos}(A_j) = \sum_{A_k | A_k \cap A_j \neq \emptyset} m(A_k) = \sum_{k=1}^{n} m(A_k) \tag{3.32}$$

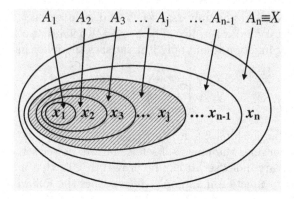

Fig. 3.6. Necessity measure $\text{Nec}(A_j)$. Sets $A_1, A_2, \ldots, A_j$ are included in set A_j.

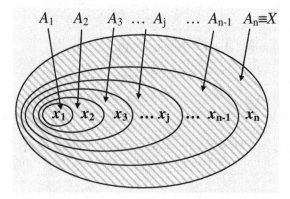

Fig. 3.7. Possibility measure $\text{Pos}(A_j)$. Sets $A_1, A_2, \ldots, A_n$ overlap with A_j.

The second part of Eqs. (3.31) and (3.32) can be readily understood by considering Fig. 3.6 and Fig. 3.7. In fact, as far as the necessity function is concerned, it is

$$A_k \subseteq A_j \quad \text{if } k \le j$$

and the necessity function is the sum of the basic probability assignment on sets A_k, with $k \le j$; on the other hand, as far as the possibility function is concerned, it is

$$A_j \cap A_k = \begin{cases} A_k & \text{if } k < j \\ A_j & \text{if } k \ge j \end{cases}$$

Hence, no intersection between the focal elements provides an empty set, and in the evaluation of the possibility function, the basic probability assignments of all sets A_k, with $k = 1, \ldots, n$, must be summed up.

If Eq. (3.8) is thus considered, it follows:

$$\text{Pos}(A_j) = 1 \tag{3.33}$$

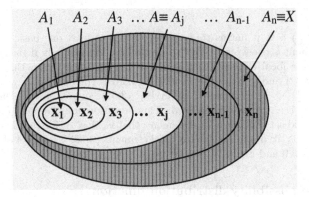

Fig. 3.8. With vertical lines: set $\bar{A}$, negation of A.

that is, the possibility function is equal to one on each (nested) focal element. Let us now consider the generic focal element A, in the complete sequence (3.20), and its negation $\bar{A}$, as reported in Fig. 3.8 with vertical lines. Let us remember that $A \cup \bar{A} = X$ and $A \cap \bar{A} = \emptyset$. The necessity and possibility functions can be evaluated for set $\bar{A}$.

As far as the necessity function is concerned, it follows that

$$\mathrm{Nec}(\bar{A}_j) = 0 \tag{3.34}$$

for every A_j, because none of the subsets $A_1, A_2, \ldots, A_n$ is included in $\bar{A}_j$.

On the other hand, as far as the possibility function is concerned, it follows that

$$\mathrm{Pos}(\bar{A}_j) = \sum_{k=j+1}^{n} m(A_k) \tag{3.35}$$

for every A_j. In fact, Fig. 3.8 shows that

$$\bar{A}_j \cap A_k \begin{cases} = \emptyset & \text{if } k \le j \\ \neq \emptyset & \text{if } k > j \end{cases}$$

and hence, only sets $A_{j+1}, A_{j+2}, \ldots, A_n$ must be considered in the evaluation of $\mathrm{Pos}(\bar{A}_j)$.

From Eqs. (3.34) and (3.35), it follows immediately that

$$\min[\mathrm{Nec}(A), \mathrm{Nec}(\bar{A})] = 0 \tag{3.36}$$

$$\max[\mathrm{Pos}(A), \mathrm{Pos}(\bar{A})] = 1 \tag{3.37}$$

In addition, necessity and possibility functions constrain each other in a strong way, as expressed by the following relationships:

$$\mathrm{Nec}(A) > 0 \Rightarrow \mathrm{Pos}(A) = 1 \tag{3.38}$$

$$\mathrm{Pos}(A) < 1 \Rightarrow \mathrm{Nec}(A) = 0 \tag{3.39}$$

Proof:

If $\text{Nec}(A) > 0$, it means that, in Eq. (3.31), at least one basic probability assignment is considered. This situation may happen only if the considered set A is a focal element; that is, $\exists i$ so that $A = A_i$. Under this situation, Eqs. (3.32) and (3.33) apply; and Eq. (3.38) is thus proved.

If $\text{Pos}(A) < 1$, it means that, in Eq. (3.32), not all focal elements A_i, for $i = 1, \ldots, n$, are considered. This situation may happen only if the considered set A is not a focal element; that is, $A \neq A_i$ for each $i = 1, \ldots, n$. Under this situation, none of the focal elements (3.20) is a subset of A; hence, $\text{Nec}(A) = 0$ and Eq. (3.39) is proved.

3.3.2 The possibility distribution function

An important property of the possibility theory is that a frame of discernment X may be completely determined by the plausibilities assigned to singletons, that is, to subsets that include only one single element x of X. In fact, given a possibility function Pos on $P(X)$, it is possible to associate it with a *possibility distribution function*:

$$r : X \to [0, 1]$$

such that

$$r(x) = \text{Pos}(\{x\}) \tag{3.40}$$

for each $x \in X$.

Thus, every possibility function Pos on $P(X)$ can be uniquely determined from the associated possibility distribution function $r : X \to [0, 1]$.

For finite universal sets, this property is expressed, for every $A \in P(X)$, by the formula[2]:

$$\text{Pos}(A) = \max_{x \in A} r(x) \tag{3.41}$$

Proof:

Relationship (3.41) can be proved by induction.

Let $|A| = 1$. Then, $A = \{x\}$, where $x \in X$. In this case, Eq. (3.41) is automatically satisfied, becoming equal to Eq. (3.40). Assume now that Eq. (3.41) is satisfied for $|A| = n - 1$ and let $A = \{x_1, x_2, \ldots, x_n\}$. Then, by Eq. (3.25),

$$\text{Pos}(A) = \max[\text{Pos}(\{x_1, x_2, \ldots, x_{n-1}\}), \text{Pos}(x_n)]$$

$$= \max[\max[\text{Pos}(x_1), \text{Pos}(x_2), \ldots, \text{Pos}(x_{n-1})], \text{Pos}(x_n)]$$

$$= \max[\text{Pos}(x_1), \text{Pos}(x_2), \ldots, \text{Pos}(x_{n-1}), \text{Pos}(x_n)] = \max_{x \in A} r(x)$$

---

[2] When X is not finite, Eq. (3.41) must be replaced with the most general equation:

$$\text{Pos}(A) = \sup_{x \in A} r(x)$$

If Eq. (3.41) is evaluated for the universal set X, the normalization condition of the possibility distribution function is readily obtained:

$$\max_{x \in X} r(x) = 1 \tag{3.42}$$

Let us also consider that the normalization condition, which applies to the universal set X, is also valid for all nested focal elements of X, as shown in Fig. 3.9.

It is possible to prove that each basic probability assignment function m represents exactly one possibility distribution function r, and vice versa. To this purpose, let us consider again the finite universal set $X = \{x_1, x_2, \ldots, x_n\}$ and assume that the focal elements are some or all of the subsets $A_i = \{x_1, x_2, \ldots, x_i\}$, for which $A_1 \subset A_2 \subset \cdots \subset A_n \equiv X$ applies. This means that, $\sum_{i=1}^{n} m(A_i) = 1$, and for every $A \neq A_i$, $m(A) = 0$.

First, it follows from Eq. (3.41) and from the definition of the possibility function as a particular plausibility function that, for every $x_i \in X$,

$$r_i = r(x_i) = \text{Pos}(\{x_i\}) = \text{Pl}(\{x_i\})$$

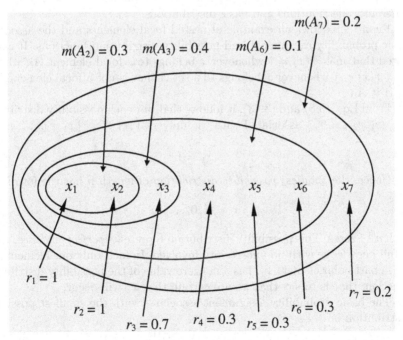

Fig. 3.9. Example of nested focal elements of the universal set X and associated basic probability assignments and possibility distribution function.

Hence, from Eqs. (3.13) and considering that any element $x_i \in X$ belongs to A_k if $k \geq i$, while $x_i \cap A_k = \emptyset$ for $k < i$, it follows that

$$r_i = \sum_{A_k|\ x_i \cap A_k \neq \emptyset} m(A_k) = \sum_{x_i \in A_k} m(A_k) = \sum_{k=i}^{n} m(A_k) = \sum_{k=i}^{n} m_k \qquad (3.43)$$

as can be also seen by Fig. 3.9. More explicitly, Eq. (3.43) can be written as follows:

$$
\begin{aligned}
r_1 &= m_1 + m_2 + \cdots + m_i + m_{i+1} + \cdots + m_n \\
r_2 &= m_2 + \cdots + m_i + m_{i+1} + \cdots + m_n \\
&\cdots \\
r_i &= m_i + m_{i+1} + \cdots + m_n \\
&\cdots \\
r_n &= \phantom{m_1 + m_2 + \cdots + m_i + m_{i+1} + \cdots +} m_n
\end{aligned}
$$

By solving these equations for m_i, the following relationship is obtained:

$$m_i = r_i - r_{i+1} \qquad (3.44)$$

where $r_{n+1} = 0$ by definition.

Equations (3.43) and (3.44) define a one-to-one correspondence between possibility distributions and basic distributions.

Figure 3.9 shows an example of nested focal elements and the associated basic probability assignment and possibility distribution functions. It can be noted that max $r(x) = 1$ whenever x belongs to a focal element A_i, whereas max $r(x) < 1$ whenever x belongs to a set A that is not a focal element; that is, $A \neq A_i$.

From Eqs. (3.8) and (3.43), it follows that, for each possibility distribution $r = \langle r_1, r_2, \ldots, r_n \rangle$ associated with the universal set $X = \{x_1, x_2, \ldots, x_n\}$, it is

$$r_1 = 1 \qquad (3.45)$$

Hence, the *smallest possibility distribution* of length n has the form:

$$r = \langle 1, 0, \ldots, 0 \rangle$$

with $n-1$ zeros. This possibility distribution represents *perfect evidence*, which is full knowledge with no uncertainty involved. In fact, only one element $x_i \in X$, in particular element x_1, has a nonzero value of the possibility distribution function; that is to say that we are certain that x_1 will occur.

The basic probability assignment associated with the smallest possibility distribution is

$$m = \langle 1, 0, \ldots, 0 \rangle$$

which has exactly the same form. In other words, the degree of belief associated with set $A_1 = \{x_1\}$ is one, and zero degrees of belief are associated with all other focal elements of X. The largest possibility distribution has the form:

$$r = \langle 1, 1, \ldots, 1 \rangle$$

with n ones. This possibility distribution represents *total ignorance*, that is, the situation in which no relevant evidence is available and the same possibility to occur is assigned to every element of X. The basic probability assignment associated with the largest possibility distribution is

$$m = \langle 0, 0, \ldots, 1 \rangle$$

The above basic probability assignment means that a zero degree of belief is associated with all focal elements except to $A_n = \{x_1, x_2, \ldots, x_n\} \equiv X$. This situation really means total ignorance, because the only available knowledge is that one among the possible elements $x_1, x_2, \ldots, x_n$ will occur and nothing more can be said. This basic probability assignment has been already introduced in the previous sections and called the vacuous belief function.

In general, it can be assessed that the larger is the possibility distribution function, the less specific is the available evidence and, consequently, the more ignorant we are.

Perfect evidence and total ignorance can be also expressed in terms of the necessity and possibility functions.

By considering Eqs. (3.31) and (3.32), it follows, for perfect evidence that

$$\text{Nec}(A_i) = 1 \quad \forall i = 1, \ldots, n$$

$$\text{Pos}(A_i) = 1 \quad \forall i = 1, \ldots, n$$

and for total ignorance that

$$\text{Nec}(A_i) = \begin{cases} 0 & \text{for } i < n \\ 1 & \text{for } i = n \end{cases}$$

$$\text{Pos}(A_i) = 1 \quad \forall i = 1, \ldots, n$$

These last equations can be also written as follows:

$$\text{Nec}(A) = \begin{cases} 0 & \text{if } A \neq X \\ 1 & \text{if } A = X \end{cases}$$

$$\text{Pos}(A) = 1 \quad \forall A$$

3.4 Fuzzy variables and possibility theory

A very interesting outcome of the mathematical formulation shown in the previous section is that the theory of the fuzzy variables can be usefully framed within the possibility theory. This approach to fuzzy variables is intuitively suggested by their representation in terms of α-cuts. In fact, the α-cuts of a fuzzy variable are a set of nested intervals, which can be assumed to be a possibilistic body of evidence.

Let us consider a fuzzy variable X and denote with X_α its generic α-cuts, with $0 \leq \alpha \leq 1$. It is immediate to verify that the α-cuts can be ordered in a nested structure. If, for instance, the values of α are limited to the finite set[3]:

$$\{0 \ 0.1 \ 0.2 \ \ldots \ 0.9 \ 1\} \tag{3.46}$$

the following applies:

$$X_{\alpha=1} \subset X_{\alpha=0.9} \subset X_{\alpha=0.8} \subset \cdots \subset X_{\alpha=0} \tag{3.47}$$

Relationship (3.47) shows that the α-cut with $\alpha = 0$ includes all other α-cuts, whereas the α-cut with $\alpha = 1$ is included by all others. The set inclusion is strict, because a fuzzy variable is, by definition, convex.[4]

As the α-cuts are ordered in a nested structure, they can be considered the focal elements of the universal set X, which corresponds to the greatest α-cut; that is, $X_{\alpha=0}$. Hence, it is possible to associate them with a basic probability assignment and, consequently, a necessity function, a possibility function, and a possibility distribution function.

In terms of α-cuts, Eqs. (3.31) and (3.32) can be rewritten as

$$\text{Nec}(X_{\alpha_j}) = \sum_{X_\alpha | X_\alpha \subseteq X_{\alpha_j}} m(X_\alpha) = \sum_{\alpha=\alpha_j}^{1} m(X_\alpha) \tag{3.48}$$

$$\text{Pos}(X_{\alpha_j}) = \sum_{X_\alpha | X_\alpha \cap X_{\alpha_j} \neq \emptyset} m(X_\alpha) = \sum_{\alpha=0}^{1} m(X_\alpha) \equiv 1 \tag{3.49}$$

Let us note that the limit of the sums in Eqs. (3.31) and (3.48), as well as those in Eqs. (3.32) and (3.49), are different. This because of the different indexing of the nested sets in Eqs. (3.21) and (3.47). In fact, in Eq. (3.21), index k takes integer values between 1 and n, and sets A_k are wider as index k increases. On the other hand, in Eq. (3.47), index α takes rational values between zero and one, and sets X are smaller as index α increases. In other

---

[3] The membership function of a fuzzy variable is a continuous function. Therefore, from a theoretical point of view, infinite α-cuts should be considered. However, if Eqs. (3.26) and (3.27) are considered, all formulations given for finite sets can be readily adapted to infinite sets.

[4] An important exception to this rule, which represents a limit case of the fuzzy variables, is given by the rectangular membership function, for which all α-cuts are equal. This membership function represents, as it will be proved later, total ignorance.

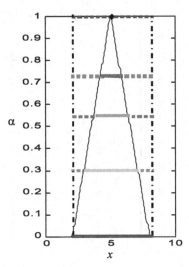

Fig. 3.10. Generic α-cuts X_α (*solid lines*) and their negations $\overline{X_\alpha}$ (*dotted lines*).

words, $k = 1$ corresponds to $\alpha = 1$; $k = 2$ corresponds to $\alpha = 0.9; \ldots$; and $k = n$ corresponds to $\alpha = 0^5$.

Let us now consider Fig. 3.10, where a set of α-cuts and their negations are shown. The negation of an α-cut X_α is a set $\overline{X_\alpha}$ that can be obtained by subtracting X_α from the universal set X, which, as stated, coincides with $X_{\alpha=0}$.

Whichever is the considered α-cut, the following relationships apply:

$$\mathrm{Nec}(\overline{X}_{\alpha_j}) \equiv 0 \tag{3.50}$$

$$\mathrm{Pos}(\overline{X}_{\alpha_j}) = 1 - \mathrm{Nec}(X_{\alpha_j}) = \sum_{\alpha=0}^{\alpha_j^-} m(X_\alpha) \tag{3.51}$$

which derive directly from Eqs. (3.34) and (3.35), respectively.

In Eq. (3.51), α_j^- denotes the value α, which precedes α_j. If, for instance, the values α are those in the finite set of Eq. (3.46), it is $\alpha_j^- = \alpha_j - 0.1$.

A fuzzy variable X can also be represented by its membership function $\mu_X(x)$. Given a value $\bar{x} \in X$, $\mu_X(\bar{x})$ quantifies, with a value between 0 and 1, to which extent element $\bar{x}$ belongs to X. According to the considerations done in the previous sections, this value can also be interpreted as the degree of belief, based on the available evidence, that element $\bar{x}$ will occur, among all possible elements in X, in other words, how much $\bar{x}$ is possible.

Therefore, it can be stated that the membership function $\mu_X(x)$ of a fuzzy variable X, defined in Chapter 2, is exactly what, in the possibility theory, is

---

[5] If 11 values are considered, as in Eq. (3.46), it is $X_{\alpha=0} \equiv A_{11}; X_{\alpha=0.1} \equiv A_{10}; X_{\alpha=0.2} \equiv A_9; \ldots$; and $X_{\alpha=1} \equiv A_1$.

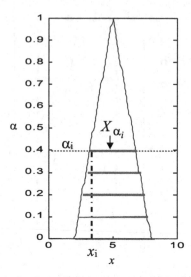

Fig. 3.11. Determination of the α-cuts to which element x_i belongs.

defined as the possibility distribution function $r_X(x)$. Formally:

$$r_X(x) = \mu_X(x) \tag{3.52}$$

for all $x \in X$.

According to the above considerations, because the smallest possibility distribution has the form $r = \langle 1, 0, \ldots, 0 \rangle$, the fuzzy variable that represents perfect evidence degenerates into a singleton; similarly, because the largest possibility distribution has the form $r = \langle 1, 1, \ldots, 1 \rangle$, the fuzzy variable that represents total ignorance is a rectangular fuzzy variable, like the one in Fig. 2.1a.

From Eq. (3.52), it follows that

$$r_i = r(x_i) = \alpha_i \tag{3.53}$$

that is, the values assumed by the possibility distribution function at the extreme points of each α-cut are equal to the considered value α, as shown in Fig. 3.11.

Of course, according to the definitions given in the previous sections, the possibility distribution function assumes different values in the different points of each α-cut. However, it is always:

$$\max_{x \in X_\alpha} r_X(x) = 1 \tag{3.54}$$

for every α.

This can also be readily understood from a geometrical point of view. Let us consider again Fig. 3.11. The peak value of the reported fuzzy variable, to which corresponds a unitary value of the membership function (and the possibility distribution function too), is given for value $x = 5$. However, by noting that this value is included in all α-cuts of the fuzzy variable, it follows that Eq. (3.54) always applies.

Then, if indexing in Eq. (3.47) is taken into account for the α-cuts, Eq. (3.43) can be rewritten as

$$r_i = \sum_{\alpha=0}^{\alpha_i} m(X_\alpha) \qquad (3.55)$$

From Eqs. (3.53) and (3.55), it follows that

$$\alpha_i = \sum_{\alpha=0}^{\alpha_i} m(X_\alpha) \qquad (3.56)$$

which more explicitly becomes

$$\alpha = 0 = m(X_{\alpha=0})$$

$$\cdots$$

$$\alpha = \alpha_j = m(X_{\alpha=0}) + \cdots + m(X_{\alpha_j})$$

$$\cdots$$

$$\alpha = 1 = m(X_{\alpha=0}) + \cdots + m(X_{\alpha_j}) + \cdots + m(X_{\alpha=1})$$

Equation (3.56) confirms that

$$\sum_{\alpha=0}^{1} m(X_\alpha) = 1$$

as required by Eq. (3.8); moreover, it is

$$m(X_{\alpha=0}) = 0$$

which proves that the α-cut $X_{\alpha=0}$ should not be considered a focal element.

Solving Eq. (3.56) for m, we obtain

$$m(X_{\alpha_j}) = \alpha_j - \alpha_j^- \qquad (3.57)$$

which is similar to (3.44), except for the different indexing.

From Eqs. (3.51), (3.56) and (3.57), the two important relationships between the possibility and necessity functions and the values of levels α follow:

$$\mathrm{Pos}(\overline{X}_{\alpha_j}) = \sum_{\alpha=0}^{\alpha_j^-} m(X_\alpha) = \sum_{\alpha=0}^{\alpha_j} m(X_\alpha) - m(X_{\alpha_j}) = \alpha_j - (\alpha_j - \alpha_j^-) = \alpha_j^-$$

$$(3.58)$$

$$\mathrm{Nec}(X_{\alpha_j}) = 1 - \mathrm{Pos}(\overline{X}_{\alpha_j}) = 1 - \alpha_j^- \qquad (3.59)$$

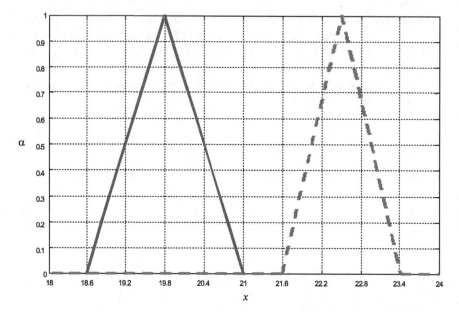

Fig. 3.12. Example of two fuzzy variables with finite support (*solid line*) and infinite support (*dashed line*).

It is important to understand now a very simple concept, which, however, could be misinterpreted.

In this book, the α-cuts of a fuzzy variable have been defined for values $0 \leq \alpha \leq 1$. However, some mathematicians prefer to define the α-cuts of a fuzzy variable only for values $0 < \alpha \leq 1$. The two formulations are obviously both valid, but it is important to understand the reasons for each of them.

The two above formulations correspond directly to the two fuzzy variables represented in Fig. 3.12. When the fuzzy variable in the solid line is considered, the value $\alpha = 0$ can be included to define the α-cuts. In fact, the universal set is considered to be a closed interval, corresponding to the greatest interval in which the possible values of the quantity could lie. According to the definitions given in this chapter and the previous one, this means that the membership function is not defined for all values of $\Re$, but only for values between 18.6 and 21 (see Fig. 3.12). In this situation, the values x for which $\mu_x \geq 0$ define an interval that exactly coincide with the universal set and that can be considered to be the α-cut at level $\alpha = 0$.

This formulation is the one chosen by the Author, because it appears to be the simpler one, as well as the more suitable for the aim of this book. For instance, in Chapter 2, fuzzy variables are used to represent measurement results, which are always included within a closed interval.

Table 3.1. Possibility and necessity functions and basic probability assignment for the α-cuts of the fuzzy variable in the solid line in Fig. 3.12 The line corresponding to $\alpha = 0$ does not refer to a focal element.

α	α-cut	$\mathrm{Pos}(X_\alpha)$	$\mathrm{Nec}(X_\alpha)$	$\mathrm{m}(X_\alpha)$	$\mathrm{Pos}(\overline{X}_\alpha)$	$\mathrm{Nec}(\overline{X}_\alpha)$
0	[18.6, 21]	1	1	0	0	0
0.1	[18.72, 20.88]	1	1	0.1	0	0
0.2	[18.84, 20.76]	1	0.9	0.1	0.1	0
0.3	[18.96, 20.64]	1	0.8	0.1	0.2	0
0.4	[19.08, 20.52]	1	0.7	0.1	0.3	0
0.5	[19.2, 20.4]	1	0.6	0.1	0.4	0
0.6	[19.32, 20.28]	1	0.5	0.1	0.5	0
0.7	[19.44, 20.16]	1	0.4	0.1	0.6	0
0.8	[19.56, 20.04]	1	0.3	0.1	0.7	0
0.9	[19.68, 19.92]	1	0.2	0.1	0.8	0
1	[19.8, 19.8]	1	0.1	0.1	0.9	0

On the other hand, when the fuzzy variable in the dashed line is considered, the value $\alpha = 0$ cannot define an α-cut. In fact, the universal set is considered to be the whole set of real numbers $\Re$. This means that the membership function is defined for all values on $\Re$, being obviously $\mu_x = 0$ for all $x \le 21.6$ and $x \ge 23.4$ (see Fig. 3.12). In this case, the values x for which $\mu_x \ge 0$ (which should correspond to the definition of the α-cut for the level $\alpha = 0$) correspond to the whole set $\Re$. Hence, they do not define a closed interval (like α-cuts require). It follows that $\alpha = 0$ cannot be considered as a possible value for the definition of an α-cut, and only values greater than zero must be taken into account.

Even if a choice has been made between the two possible formulations, it is important to understand that they lead to the same results. In fact, in one case, the α-cut at level $\alpha = 0$ does not exist, whereas in the other case, it exists but does not represent a focal element, as previously proved. Hence, focal elements of a fuzzy variable are in both cases defined for levels $0 < \alpha \le 1$.

Let us now consider, as an example, the fuzzy variable in the solid line of Fig. 3.12. Let, for the sake of simplicity, the values of α be limited to the finite set $\{0\ 0.1\ 0.2\ \ldots\ 0.9\ 1\}$. Table 3.1 shows the values of the necessity and possibility functions for each α-cut and its negation, and the values of the basic probability assignment function.

From this table, the following considerations can be drawn, which confirm the theoretical considerations:

- interval [18.6, 21] is not a focal element. In fact, the basic probability assignment is zero;
- except for the α-cut at level $\alpha = 0$, the basic probability assignment is equal for all α-cuts. However, this is only a consequence of having chosen equally spaced α-levels;

- the necessity function increases as the width of the α-cuts increases, that is, as α decreases. This is coherent with the mathematical theory, as well as with the consideration that the total belief about one set is greater as the set is greater;
- the possibility function is 1 for all α-cuts;
- the possibility function of the negation of the α-cuts increases as α increases. This is coherent with the fact that, as the width of the α-cut decreases, the total belief about that α-cut itself decreases and the possibility that something outside the α-cut occurs is greater;
- the necessity function is zero outside the α-cuts.

In the previous discussion, in order to represent a fuzzy variable, a finite number of values α has been considered. However, because the membership function of a fuzzy variable is a continuous function, from a theoretical point of view, an infinite number of levels α should be considered. Again from a theoretical point of view, this can be done if Eqs. (3.26) and (3.27) are employed to define the necessity and possibility functions, respectively. The mathematical relationship derived above can be reformulated according to these definitions. The most interesting result of this modified formulation, valid for continuous membership functions, is that Eq. (3.59) becomes, for a given α that varies continuously between 0 and 1:

$$\text{Nec}(X_{\hat{\alpha}}) = 1 - \hat{\alpha} \qquad (3.60)$$

This result can also be intuitively achieved because $\lim_{n \to \infty}(\alpha_j - \alpha_j^-) = 0$, and therefore, as the number of α-cuts employed to represent the fuzzy variable tends to infinity, Eq. (3.59) tends to Eq. (3.60).

In practical applications, membership functions are described in a discrete form, for which Eq. (3.59) should be applied. However, if the number of levels is high enough, the difference between Eqs. (3.59) and (3.60) becomes insignificant, and this last one can be used for the sake of simplicity.

3.5 Probability theory

The probability theory is a well-known and widely used mathematical theory, which deals with random phenomena. What is generally ignored is that the probability theory can be defined within the Theory of Evidence, by applying a suitable constraint, different from the one applied to derive the possibility theory, to the bodies of evidence.

As already reported at the beginning of this chapter, a probability function Pro is required to satisfy the following equation:

$$\text{Pro}(A \cup B) = \text{Pro}(A) + \text{Pro}(B) \qquad (3.61)$$

for all sets A and $B \in P(X)$ such that $A \cap B = \emptyset$.

This requirement, which is usually referred to as the additivity axiom of probability functions, is stronger than the one given in Eq. (3.4) for belief functions. This implies that the probability functions are a special type of belief functions. The relationship between belief and probability functions is given by the following theorem:

Theorem 3.1. *A belief function* Bel *on a finite power set* $P(X)$ *is a probability function if and only if the associated basic probability assignment function* m *is given by* $m(\{x\}) = \text{Bel}(\{x\})$ *and* $m(A) = 0$ *for all subsets of* X *that are not singletons.*

Proof:

- Assume that Bel is a probability function. For the empty set $\emptyset$, the theorem trivially holds, because $m(\emptyset) = 0$ by definition of m.
  Let $A \neq \emptyset$, and assume $A = \{x_1, x_2, \ldots, x_n\}$. Then, by repeated application of the additivity axiom in Eq. (3.61), we obtain

  $$\text{Bel}(A) = \text{Bel}(\{x_1\}) + \text{Bel}(\{x_2, x_3, \ldots, x_n\})$$
  $$= \text{Bel}(\{x_1\}) + \text{Bel}(\{x_2\}) + \text{Bel}(\{x_3, x_4, \ldots, x_n\})$$
  $$= \cdots = \text{Bel}(\{x_1\}) + \text{Bel}(\{x_2\}) + \cdots + \text{Bel}(\{x_n\})$$

  As $\text{Bel}(\{x_1\}) = m(\{x\})$ for any $x \in X$ by (3.10), we have

  $$\text{Bel}(A) = \sum_{i=1}^{n} m(\{x_i\})$$

  Hence, Bel is defined in terms of a basic assignment that focuses only on singletons.

- Assume now that a basic probability assignment function m is given such that

  $$\sum_{x \in X} m(x) = 1$$

  Then, for any sets $A, B \in P(X)$ such that $A \cap B = \emptyset$, we have

  $$\text{Bel}(A) + \text{Bel}(B) = \sum_{x \in A} m(x) + \sum_{x \in B} m(x)$$
  $$= \sum_{x \in A \cup B} m(x) = \text{Bel}(A) \cup \text{Bel}(B)$$

  Consequently, Bel is a probability function. This completes the proof.

According to this theorem, probability functions on finite sets are fully represented by a function:

$$p : X \to [0, 1]$$

such that

$$p(x) = m(\{x\}) \tag{3.62}$$

This function p is usually called the *probability distribution function*, whereas set $\mathbf{p} = \langle p(x) | x \in X \rangle$ is called the *probability distribution* on X.

When the basic probability assignment focuses only on singletons, as required by probability functions, the right-hand sides of Eqs. (3.10) and (3.13) assume the same value for every $A \in P(X)$; that is,

$$\text{Bel}(A) = \text{Pl}(A) = \sum_{x \in A} m(\{x\})$$

In fact, because focal elements are singletons, the focal elements included in set A, as required by Eq. (3.10), are all elements of set A itself; similarly, the focal elements whose intersection with set A is not an empty set, as required by Eq. (3.13), are again all elements of set A itself.

The previous equation can be also written, by taking into account Eq. (3.62), as follows:

$$\text{Bel}(A) = \text{Pl}(A) = \sum_{x \in A} p(x)$$

Hence, it can be stated that belief and plausibility functions merge when the focal elements are singletons and become the well-known probability function:

$$\text{Pro}(A) = \sum_{x \in A} p(x) \tag{3.63}$$

From Eqs. (3.8), (3.62), and (3.63), the normalization condition for probability distribution functions follows:

$$\sum_{x \in X} p(x) = 1 \tag{3.64}$$

Within the probability theory, total ignorance over the universal set X is expressed by the uniform probability distribution function:

$$p(x) = \frac{1}{|X|} \tag{3.65}$$

for every $x \in X$.

This formulation can be explained in a similar way as done, at the beginning of this chapter, for the Bayesian belief functions. Let us consider the simple example where the universal set X contains only two elements, say x_A and x_B. Then, because of Eqs. (3.61) and (3.64), it is

$$1 = \text{Pro}(x_A) + \text{Pro}(x_B)$$

Because $\text{Pro}(x) = p(x)$ for any $x \in X$ by (3.63), it is also

$$1 = p(x_A) + p(x_B)$$

In case of total ignorance, there is no reason to assign a greater value of the probability distribution to one element, rather than to the other one. Hence, the probability distribution that expresses total ignorance in the case of a two-element universal set is

$$\left\langle \frac{1}{2}, \frac{1}{2} \right\rangle$$

Equation (3.65) is readily proven in a similar way, when the universal set contains more than two elements. In case the universal set is a closed interval, containing an infinite number of elements, $|X|$ in Eq. (3.65) must be interpreted as the width of the interval itself.

The probability theory is a well-assessed and known mathematical theory, and the properties of the probability functions are well known. Therefore, nothing else is added here to the given proof that the probability theory can be formulated as a particular case of the Theory of Evidence, because this is the only point of deep interest for the rest of the book.

The basic and typical concepts of the probability theory that are required to understand the mathematical formulations reported in the following sections of this book are hence supposed to be familiar to the reader.

3.5.1 Comparison between possibility and probability theories

The purpose of this section is to compare probability and possibility theories. It has been shown, through this chapter, that probability theory and possibility theory are distinct theories. Hence, the best way to compare them is to examine the two theories from a broader perspective. Such a perspective is offered by the Theory of Evidence, within which probability and possibility theories are seen as special branches.

To facilitate the comparison, the basic mathematical properties of both theories are summarized in Table 3.2, whereas Fig. 3.13 shows the relationships between all functions that has been defined in this chapter: belief and plausibility functions, defined in the more general Theory of Evidence; necessity and possibility functions, defined within the possibility theory; and probability function, defined within the probability theory.

As shown in Fig. 3.13, probability, necessity and possibility functions do not overlap with one another, except for one point. This point represents a very special situation: *perfect evidence*. Perfect evidence is characterized by a crisp value, to which a total belief is associated. Hence, in the Theory of Evidence, this situation is characterized by a universal set with only one focal element; moreover, this focal element is a singleton. As obvious by their mathematical properties, in this particular case, the probability distribution function and the possibility distribution function become equal:

$$p(\bar{x}) = r(\bar{x}) = 1$$

$$p(x) = r(x) = 0 \quad \text{for every} \quad x \neq \bar{x}$$

where $\bar{x}$ is the single focal element.

Table 3.2. Probability and possibility theories: Comparison of the mathematical properties for finite sets.

	Probability theory	Possibility theory
BODIES OF EVIDENCE	Singletons	Family of nested sets
FUNCTIONS DEFINED IN THE THEORY	Probability function, Pro	Possibility function, Pos Necessity function, Nec
RULES	Additivity: $\text{Pro}(A \cup B) =$ $= \text{Pro}(A) + \text{Pro}(B) +$ $-\text{Pro}(A \cap B)$	Max/Min rules: $\text{Nec}(A \cap B) =$ $= \min[\text{Nec}(A), \text{Nec}(B)]$ $\text{Pos}(A \cup B) =$ $= \max[\text{Pos}(A), \text{Pos}(B)]$
RELATIONSHIPS	Not applicable	$\text{Nec}(A) = 1 - \text{Pos}(\overline{A})$ $\text{Pos}(A) < 1 \Rightarrow \text{Nec}(A) = 0$ $\text{Nec}(A) > 0 \Rightarrow \text{Pos}(A) = 1$
RELATIONSHIPS	$\text{Pro}(A) + \text{Pro}(\overline{A}) = 1$	$\text{Pos}(A) + \text{Pos}(\overline{A}) \geq 1$ $\text{Nec}(A) + \text{Nec}(\overline{A}) \leq 1$ $\max[\text{Pos}(A), \text{Pos}(\overline{A})] = 1$ $\min[\text{Nec}(A), \text{Nec}(\overline{A})] = 0$
DISTRIBUTION FUNCTIONS DEFINED IN THE THEORY	Probability distribution function $p : X \rightarrow [0, 1]$ $\text{Pro}(A) = \sum_{x \in A} p(x)$	Possibility distribution function $r : X \rightarrow [0, 1]$ $\text{Pos}(A) = \max_{x \in A} r(x)$
NORMALIZATION CONDITION	$\sum_{x \in X} p(x) = 1$	$\max_{x \in X} r(x) = 1$
EXPRESSION OF TOTAL IGNORANCE	$p(x) = \dfrac{1}{|X|} \ \forall x \in X$	$r(x) = 1 \ \forall x \in X$

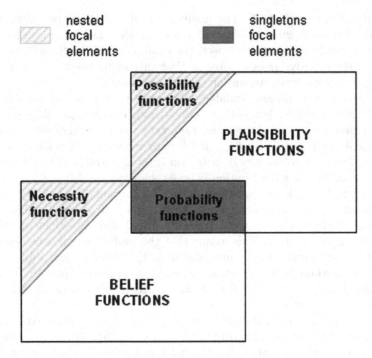

Fig. 3.13. Inclusion relationship among belief, plausibility, probability, necessity, and possibility functions.

However, the probability distribution function $p(x)$ and the possibility distribution function $r(x)$ assume the same value only in this particular case. In fact, they are defined in a very different way and they satisfy a different normalization condition, as also shown in Table 3.2. Moreover, they allow us to express total ignorance in a very different way, as already underlined throughout this chapter.

This is a very crucial point, because it allows us to understand in which situations one theory is preferable than the other one, and vice versa.

In the possibility theory, total ignorance is expressed as follows:

$$r(x) = 1 \quad \text{for every} \quad x \in X$$

where X is the universal set.

On the contrary, in the probability theory, total ignorance is expressed by the uniform probability distribution on the universal set:

$$p(x) = \frac{1}{|X|} \quad \text{for every} \quad x \in X$$

Hence, because no evidence is available that one focal element is preferable to the others, an equal probability value is assigned to all focal elements of the considered universal set X. The justification of this choice has already been discussed at the beginning of this chapter, together with the inconsistency of it (let us remember, in this respect, the example of life on Sirius). Moreover, also purely intuitive reasons suggest that this requirement is too strong to obtain an honest characterization of total ignorance.

In fact, if the unique available evidence is that a certain number of elements are plausible, but nothing else is said, it is not possible to guess a certain probability distribution function, rather than another one. In fact, no probability distribution is supported by any evidence. Therefore, the choice of one particular probability distribution is totally arbitrary, because there is no evidence that one distribution is preferable over another.

Therefore, if it is not possible to choose a particular probability distribution function, total ignorance should be expressed, in the probability theory, in terms of the full set of the probability distribution functions on the considered universal set X. This means that the probability of each element of X should be allowed to take any value in $[0, 1]$. However, such a formulation, which is based on imprecise probabilities, is foreign to the probability theory and the uniform probability distribution is erroneously universally adopted to express total ignorance.

Hence, total ignorance should not be mathematically expressed by a probability distribution function, but only by a possibility distribution function. This also means that only a fuzzy variable can suitably represent total ignorance.

These differences in the mathematical properties of the probability and possibility distribution functions make each of the two theories (probability and possibility theories) more suitable for modeling certain types of incomplete knowledge and less suitable for modeling other types. Probability theory, with its probability distribution functions, is the ideal tool to represent incomplete knowledge in those situations where class frequencies are known or where evidence is based on outcomes of a sufficiently long series of independent random experiments. In fact, only in this case can a probability distribution function be assigned to the given universal set. On the other hand, the possibility theory, with its possibility distribution functions, is the ideal tool to formalize incomplete knowledge expressed in terms of fuzzy propositions, or, in other words, in terms of closed sets. That is to say, probability distribution functions can only represent random phenomena, as widely known, whereas fuzzy variables can represents unknown systematic phenomena and, as widely explained in this chapter, total ignorance.

Relationships between probability, possibility and necessity functions can also be obtained if a different interpretation is considered.

In fact, belief and plausibility functions may be interpreted as lower and upper probability estimates [KY95]; that is, for every set $A \in P(X)$, the two dual functions Bel and Pl form an interval $[\text{Bel}(A), \text{Pl}(A)]$, which is interpreted as imprecise estimates of probabilities. Also, the possibility theory can be interpreted in terms of imprecise probabilities, provided that the normalization requirement is applied. Due to the nested structure of the focal elements, the intervals of estimated probabilities are not totally arbitrary and become:

- $[0, \mathrm{Pos}(A)]$ if $\mathrm{Pos}(A) < 1$
- $[\mathrm{Nec}(A), 1]$ if $\mathrm{Nec}(A) > 0$

When the probability theory is considered, of course, the intervals of estimated probabilities degenerate into a crisp value: $\mathrm{Pro}(A)$.

Although the interpretations of the possibility theory are still less developed than their probabilistic counterparts, viewing necessity and possibility functions as lower and upper probabilities opens a bridge between the possibility theory and the probability theory [DP93, KH94, R89].

When information regarding some phenomena is given in both probabilistic and possibilistic terms, the two descriptions should be consistent. That is, given a probability function, Pro, and a possibility function, Pos, both defined on $P(X)$, the two functions must satisfy some consistency conditions. Various consistency conditions may be required. The weakest one is expressed, for every $A \in P(X)$, as follows: An event that is probable to some degree must be possible at least to the same degree. Formally:

$$\mathrm{Pro}(A) \leq \mathrm{Pos}(A)$$

for all $A \in P(X)$. On the contrary, the strongest consistency condition requires that any event with nonzero probability must be fully possible. Formally:

$$\mathrm{Pro}(A) > 0 \Rightarrow \mathrm{Pos}(A) = 1$$

for all $A \in P(X)$. Of course, other consistency conditions may also be formulated that are stronger than the first one and weaker than the second one.

The consistency between probability and possibility measures, at least in its weakest form, is an essential requirement in any probability–possibility transformation. The motivation of studying probability–possibility transformations arises not only from a desire to comprehend the relationship between the two theories, but also from some practical problems: constructing a membership function of a fuzzy variable from statistical data, combining probabilistic and possibilistic information and transforming probabilities in possibilities to reduce computational complexity.

An original probability–possibility transformation is discussed in Chapter 5.

4

Random–Fuzzy Variables

In the previous chapter, the fundamentals of the possibility and probability theories have been recalled within the more general framework of the Theory of Evidence, which embraces both of them. The different ways with which the two theories represent incomplete knowledge have also been discussed and compared.

A very interesting aspect of the possibility theory is that fuzzy variables can be defined within this theory and can be considered as the natural variables of the possibility theory; in a similar way, the random variables are considered the natural variables of the probability theory.

Of course, random variables and fuzzy variables are different mathematical objects, defined within two distinct mathematical theories. In particular, it has been proven that a random variable is described by its probability distribution function $p(x)$, whereas a fuzzy variable is described by its possibility distribution function $r(x)$. Although, at a first sight, these distributions could appear similar to each other and they seem to handle incomplete knowledge in a similar way, they represent different concepts and show a significant difference, strictly related to the different normalization conditions to which they obey: $p(x)$ has an unitary subtended area, whereas the maximum value of $r(x)$ is 1.

It has been shown that, due to the different ways the probability and possibility distribution functions have been defined, and due to the different bodies of evidence they are based on, random variables and fuzzy variables describe different situations of incomplete knowledge, and therefore, they never provide the same kind of information. There is only one exception to this situation, when total knowledge is achieved. In fact, in case of total knowledge, the available evidence shows, with full certainty, that element x takes value x_0; that is, $x = x_0(x_0 \in X)$. Hence, it is

$$p(x)|_{x=x_0} = r(x)|_{x=x_0} = 1$$
$$p(x)|_{x \neq x_0} = r(x)|_{x \neq x_0} = 0$$

This is the only particular case in which random and fuzzy variables degenerate into the same crisp value $x = x_0$.

According to the mathematical theory developed in the previous chapters, the main difference between the probability and possibility theories lies in the different characteristics of their bodies of evidence and in the fact that the probability theory satisfies Bayes additivity condition, whereas the possibility theory does not.

The body of evidence to which the probability theory refers is a set of singletons x, for which $p(x) \neq 0$. This means that, on the basis of the available evidence, $p(x_0)$ quantifies the probability that an event $x \in X$ occurs in x_0. Therefore, $p(x)$ represents the probability on how single events x distribute in the universal set X.

According to the above considerations, the probability theory is particularly suitable to quantify incomplete knowledge when the reasons for incompleteness lie in the presence of random phenomena that can be modeled by a proper probability distribution function $p(x)$ and that force the occurrence of a given event x to distribute randomly on X, according to the same function $p(x)$. We can then represent the result of a measurement with a random variable and its associated probability distribution function only if random effects are the sole effects responsible for measurement uncertainty.

On the other hand, the body of evidence to which the possibility theory refers is a set of nested sets, for which the associated basic probability assignment function is nonzero. It has been shown that the basic probability assignment function $m(A)$, which is directly and uniquely related to the possibility distribution function $r(x)$ of each element $x \in A$, quantifies the belief that an event $x \in X$ occurs in A. However, from the knowledge of $m(A)$, that is the degree of belief associated with set A, it is not possible to quantify the degree of belief in the subsets of A: In other words, nothing can be said about where, in A, the event occurs.

According to these considerations, the possibility theory is particularly suitable to quantify incomplete knowledge when the reasons for incompleteness lie in the presence of phenomena that are known to cause the occurrence of a given event x to fall inside A, but that are not known enough to make any additional assumption about where, in A, the event falls. Therefore, when we deal with measurement results, we can represent them within the possibility theory, and therefore, with a fuzzy variable, whenever unknown phenomena, or uncompensated systematic phenomena, are supposed to be responsible for measurement uncertainty, because, in this case, the amount of available information fits perfectly with the one described above.

These theoretical considerations can be better understood if the example in Fig. 4.1 is taken into account. Two generic measurement results R_1 and R_2, which are supposed to be distributed according to a Gaussian function, are considered and their mean value is evaluated.

In Fig. 4.1a, the two results (solid line) are represented in terms of possibility distribution functions, whereas in Fig. 4.1b, they are represented in

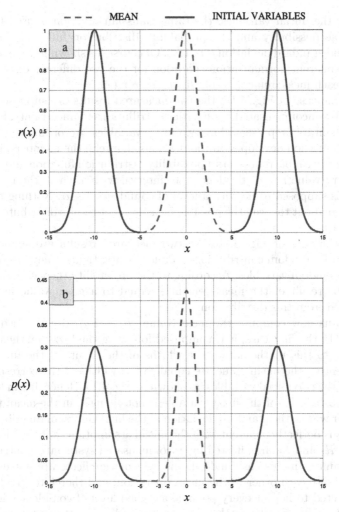

Fig. 4.1. (a) Mean of two fuzzy variables; (b) mean of two random variables.

terms of probability distribution functions (solid line). This example is aimed
at showing the different behavior of fuzzy and random variables when they
are composed to evaluate their mean value. For this reason, the same shape of
the variables has been considered, although this does not automatically mean
that they carry the same kind of information.[1]

---

[1] It will be shown in Chapter 5 that it is possible to apply suitable probability–
possibility transformations, in order to obtain a possibility distribution that represents information equivalent to that represented by a probability distribution.
In this case, however, the two distributions show, generally, different shapes.

For the initial variables, the same Gaussian function is used to describe both the possibility and the probability distribution functions: mean value ± 10 and standard deviation $\sigma = 1.33$. Of course, as shown in Fig. 4.1, because of the different normalization condition for the two different distributions, it is, for each measurement result R_i: $p_i(x) \neq r_i(x)$, $\forall x$.

In the case of Fig. 4.1a, the two measured results are supposed to be affected by uncompensated systematic contributions to uncertainty. Hence, they can be suitably represented by two fuzzy variables. According to the theoretical formulation developed in the previous chapters, the membership function of a fuzzy variable represents a possibility distribution function and allows one, for every value $r = \alpha$, to identify an interval within which the measurement result is supposed to lie with a level of confidence $1 - \alpha$. Nothing else can be assumed about the way the possible measurement results distribute within the interval.

In the case of Fig. 4.1b, the two measured results are supposed to be affected by random contributions to uncertainty. Hence, they are represented by two random variables. According to the probability theory, the probability that the result of the measurement is equal to a given value in X can be obtained from this distribution.

Let us now compute the mean value $(R_1 + R_2)/2$ of the two measured values. In the first case, it is computed following the fuzzy mathematics. According to this mathematics, the widths of the α-cuts of the fuzzy variable representing the mean value are equal to the widths of the corresponding α-cuts of the two initial variables, as shown in Fig. 4.1a. Hence, it can be assessed that the "width" of the fuzzy variable is not changed in the mean process, or in other words, the standard deviation of the final Gaussian function is exactly σ. This is perfectly coherent with the initial assumption that the measurement results R_1 and R_2 are affected by uncompensated systematic contributions to uncertainty: In this case, indeed, the systematic effects are not expected to compensate each other in the mean operation, and therefore, the final result is expected to lie, for every possible assigned level of confidence, in intervals with the same widths as the ones corresponding, in the initial variables, to the same level of confidence.

In the second case, when the measurement results are represented by random variables, statistics must be applied. Taking into account that the mean of two Gaussian distributions with standard deviations σ is still a Gaussian distribution, with standard deviation equal to $\sigma/\sqrt{2}$, the mean value shows the probability distribution reported in Fig. 4.1b (dashed line). Again, this is perfectly coherent with the initial assumption that the measurement results are affected by random contributions to uncertainty: In this case, the random effects are supposed to partially compensate each other in the mean operation, providing a narrower distribution of the final result.

The above considerations, together with this simple example, show that the distribution of the possible values that can be reasonably assigned to the result of a measurement can be described in terms of a probability distribution (and hence framed within the probability theory) only if the contributions to uncertainty are random ones. On the other hand, the distribution of the possible values that can be reasonably assigned to the result of a measurement can be described in terms of a possibility distribution (and hence framed within the possibility theory) only if the contributions to uncertainty are not random, that is, for example, total ignorance and unknown, or uncompensated, systematic contributions.

The problem is that, in most practical applications, all different kinds of contributions to uncertainty are affecting the measurement result and must be then considered together.

Of course, referring to two different theories, although perfectly coherent from a strict theoretical point of view, would be impractical. The aim of this chapter is to solve this problem and find a single mathematical object able to consider these contributions.

Because of the mathematical relationship among probability theory, possibility theory, and Theory of Evidence, it seems natural to look for this mathematical object within the Theory of Evidence, which encompasses the two former ones as particular cases. This object is the *random–fuzzy variable*. Random–fuzzy variables (RFVs) can be defined as particular cases of fuzzy variables of type 2, which are already known mathematical objects [KG91]; though, however, they have never been used for representing and processing uncertainty in measurement.

This chapter shows how RFVs can be defined to cope with this problem, and the next chapters will define an original mathematics for the RFVs.

4.1 Definition of fuzzy variables of type 2

Before giving the definition of a random–fuzzy variable, it is necessary to introduce the concept of interval of confidence of type 2 and fuzzy number of type 2.

It is known that an interval of confidence is a closed interval in $\Re$, within which it is possible to locate the possible values of an uncertain result. For instance,

$$A = [a_1, a_2] \tag{4.1}$$

as represented in Fig. 4.2a.

As shown in Chapter 3, a set of confidence intervals of type 1:

$$A_\alpha = [a_1^\alpha, a_2^\alpha] \tag{4.2}$$

which depend on a value α, with $\alpha \in [0,1]$, and adhere to the following constraint:

$$\alpha' > \alpha \;\Rightarrow\; A_{\alpha'} \subset A_\alpha$$

defines the membership function of a fuzzy variable. In other words, the intervals of confidence of type 1 (4.2) are indeed the α-cuts defined in Eq. (2.1).

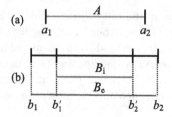

Fig. 4.2. (a) Interval of confidence of type 1; (b) interval of confidence of type 2.

Let us now assume that the lower and upper bounds of the interval of confidence (4.1) are themselves uncertain; hence, instead of being ordinary numbers, they are confidence intervals; that is,

$$B = [\,[b_1, b_1'], [b_2', b_2]\,] \qquad (4.3)$$

where $b_1 \leq b_1' \leq b_2' \leq b_2$. This means that the left bound of the confidence interval can vary between b_1 and b_1', and the right bound can vary between b_2 and b_2', as shown in Fig. 4.2b. When $b_1 = b_1'$ and $b_2 = b_2'$, the interval of confidence of type 2 becomes an interval of type 1; if $b_1 = b_1' = b_2 = b_2'$, the interval of confidence of type 0 is obtained, that is, an ordinary number.

The properties of intervals of confidence of type 2 are similar to those of intervals of confidence of type 1, reported in Section 2.2. Moreover, they also obey to interval analysis.

In this respect, it is useful to consider intervals of confidence of type 2 as the set of two intervals of confidence of type 1:

$$B_i = [b_1', b_2'] \text{ and } B_e = [b_1, b_2]$$

where B_i refers to the internal interval and B_e refers to the external one, as reported in Fig. 4.2b. Hence, it can be stated that, given two intervals of confidence of type 2:

$$B = [\,[b_1, b_1'], [b_2', b_2]\,]$$
$$C = [\,[c_1, c_1'], [c_2', c_2]\,]$$

and the generic arithmetic operation $B * C$, where $*$ can be $+, -, \times$, and $\div$, it is

$$(B * C)_i = B_i * C_i$$
$$(B * C)_e = B_e * C_e$$

where $B_i * C_i$ and $B_e * C_e$ are evaluated according to the mathematics of the intervals reported in Section 2.2.

An alternative, simpler notation than the one in Eq. (4.3) to refer to intervals of confidence of type 2 is

$$B = \{b_1, b_2, b_3, b_4\} \tag{4.4}$$

In the following, this notation will be used for the sake of simplicity.

Similarly to intervals of confidence of type 1, which can be used to define a fuzzy variable, intervals of confidence of type 2 can be used to define fuzzy variables of type 2. In this respect, let us consider a sequence of intervals of confidence of type 2, which depend on a value α, where $\alpha \in [0, 1]$:

$$B^\alpha = \{b_1^\alpha, b_2^\alpha, b_3^\alpha, b_4^\alpha\}$$

Let us also give the following constraints:

- $b_1^\alpha \le b_2^\alpha \le b_3^\alpha \le b_4^\alpha$, $\forall \alpha$;
- the sequence of intervals of confidence of type 1 $[b_1^\alpha, b_4^\alpha]$ generates a membership function that is normal and convex;
- the sequence of intervals of confidence of type 1 $[b_2^\alpha, b_3^\alpha]$ generates a membership function that is convex;
- $\forall \alpha, \alpha'$:

$$\alpha' > \alpha \Rightarrow \begin{cases} [b_1^{\alpha'}, b_3^{\alpha'}] \subset [b_1^\alpha, b_3^\alpha] \\ [b_2^{\alpha'}, b_4^{\alpha'}] \subset [b_2^\alpha, b_4^\alpha] \end{cases} \quad \text{for } b_2^{\alpha'} \le b_3^{\alpha'} \tag{4.5}$$

- if the maximum of the membership function generated by the sequence of confidence intervals $[b_2^\alpha, b_3^\alpha]$ is found at level α_m, then

$$[b_2^{\alpha_m}, b_3^{\alpha_m}] \subseteq [b_1^{\alpha=1}, b_4^{\alpha=1}] \tag{4.6}$$

In this case, the sequence of intervals of confidence of type 2 describes a fuzzy number of type 2. Examples of fuzzy numbers of type 2 are shown in Fig. 4.3. Moreover, Fig. 4.4 illustrates properties (4.5) and (4.6).

Particular cases of fuzzy variables of type 2 can be defined by adding more restrictive constraints to the sequence of intervals $B^\alpha = \{b_1^\alpha, b_2^\alpha, b_3^\alpha, b_4^\alpha\}$, where $\alpha \in [0, 1]$.

For instance, in a fuzzy variable of type 2, it is not required that the internal membership function is normal; hence, a particular case is given by the fuzzy variables of type 2 where both membership functions, generated by $[b_1^\alpha, b_4^\alpha]$ and $[b_2^\alpha, b_3^\alpha]$, are convex and normal. An example is shown in Fig. 4.3a.

This kind of fuzzy variables of type 2 satisfies the following constraints:

- $b_1^\alpha \le b_2^\alpha \le b_3^\alpha \le b_4^\alpha$, $\forall \alpha$;
- the sequences of intervals of confidence of type 1 $[b_1^\alpha, b_4^\alpha]$ and $[b_2^\alpha, b_3^\alpha]$ generate membership functions that are normal and convex;

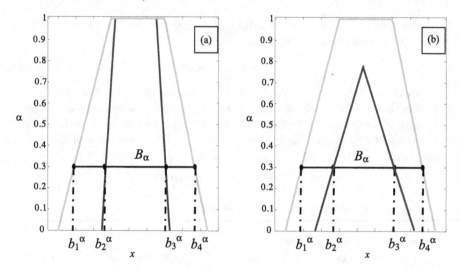

Fig. 4.3. Examples of fuzzy numbers of type 2.

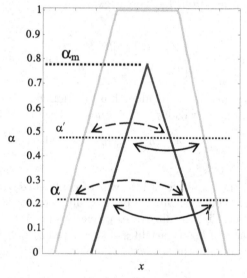

Fig. 4.4. Properties (4.5) and (4.6) for a fuzzy number of type 2. Let us consider the intervals indicated by the equal arrows. It is easily understood that intervals corresponding to a greater value α are included in those corresponding to smaller values; the maximum of the internal membership function is obtained for value α_m.

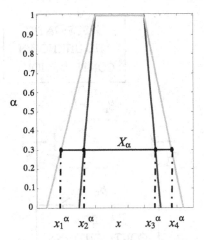

Fig. 4.5. Examples of a particular fuzzy variable of type 2.

• $\forall \alpha, \alpha'$:

$$\alpha' > \alpha \Rightarrow \begin{cases} [b_1^{\alpha'}, b_3^{\alpha'}] \subset [b_1^{\alpha}, b_3^{\alpha}] \\ [b_2^{\alpha'}, b_4^{\alpha'}] \subset [b_2^{\alpha}, b_4^{\alpha}] \end{cases}$$

•

$$[b_2^{\alpha=1}, b_3^{\alpha=1}] \subseteq [b_1^{\alpha=1}, b_4^{\alpha=1}] \qquad (4.7)$$

where the second and the fourth constraints are stricter than those used to define a general fuzzy number of type 2.

The last condition requires that interval $[b_2^{\alpha=1}, b_3^{\alpha=1}]$, which is derived from the internal membership function at level $\alpha = 1$, is included in interval $[b_1^{\alpha=1}, b_4^{\alpha=1}]$, which is derived from the external membership function at the same level α. Moreover, another constraint can be added; that is,

$$[b_2^{\alpha=1}, b_3^{\alpha=1}] \equiv [b_1^{\alpha=1}, b_4^{\alpha=1}] \qquad (4.8)$$

In this case, intervals $[b_2^{\alpha=1}, b_3^{\alpha=1}]$ and $[b_1^{\alpha=1}, b_4^{\alpha=1}]$ must always coincide. An example is shown in Fig. 4.5.

4.1.1 The random–fuzzy variables

In the previous section, the fuzzy variables of type 2 have been defined starting from the concept of intervals of confidence of type 2. As stated, an interval of confidence of type 2 is a confidence interval for which the bounds are uncertain. This means that the lower and upper bounds are not defined by crisp values but only by given intervals, within which the bounds themselves can lie. Hence, by definition, also these intervals are confidence intervals. This means that the width of the confidence interval B described by Eq. (4.4) can vary between a

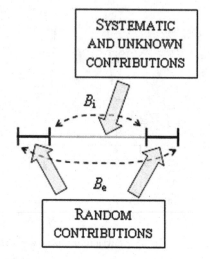

Fig. 4.6. Meaning of the different parts of an interval of confidence of type 2.

minimum value, given by the width of the internal interval B_i, and a maximum value, given by the width of the external interval B_e (see Fig. 4.2b).

The internal interval B_i is, at all effects, an interval of confidence of type 1. As shown, a sequence of confidence intervals of type 1 defines a fuzzy variable. Moreover, in Chapter 2, it has been shown that fuzzy variables are suitable to represent measurement results affected by systematic contributions to uncertainty, or by contributions whose nature is unknown (total ignorance).

The same can be of course applied to the external interval B_e. However, because B_i and B_e are not defined separately, but define the minimum and maximum variations of the given confidence interval B, it is possible to associate a random nature with this variation. In other words, it can be intuitively stated that the random contributions to uncertainty are represented by the side intervals of each interval of confidence of type 2. This is shown in Fig. 4.6.

According to the above assumption, which is perfectly coherent with the general definition of a confidence interval of type 2, a fuzzy variable of type 2 can be intuitively considered as a suitable tool for representing a measurement result in every situation, that is, whichever is the nature of the contributions affecting the result itself. In fact, by definition, a fuzzy variable of type 2 is a set of confidence intervals of type 2, or, equivalently, is a set of α-cuts.

Of course, the above intuitive statements can be turned into a stricter approach only if the two following requirements are met:

- The internal membership function of the fuzzy variable of type 2 must be built in accordance with the available information regarding the systematic or unknown contributions affecting the measurement result. On the other hand, the external membership function of the fuzzy variable of type 2 must be built in accordance with the probability distribution function $p(x)$, which describes the distribution of the random contributions to uncertainty. In other words, the probability distribution function $p(x)$ must be transformed into an equivalent possibility distribution function $r(x)$. This of course requires the definition of a suitable probability–possibility transformation, as dealt with in Chapter 5.

• When fuzzy variables of type 2 must be mathematically combined, a different behavior must be considered for the internal and the external parts. In fact, for each fuzzy variable of type 2, the internal part represents a fuzzy variable of type 1 and hence must adhere to the mathematics defined for these variables. This means that, when two fuzzy variables of type 2 are combined, the corresponding confidence intervals B_i of each variable are combined according to the mathematics of the intervals. On the other hand, the external part of fuzzy variables of type 2 represents the variations of each confidence interval B_i due to random effects. Therefore, the behavior of the random parts must be similar to the behavior of the probability distribution functions when they combine together. To achieve this requirement, a suitable mathematics must be defined, as reported in Chapter 7.

Furthermore, the result of a measurement cannot be represented by a totally generic fuzzy variable of type 2 but only by a particular subclass. In fact, let us consider the following points.

If the measurement result is affected by different kinds of uncertainty contributions (random and nonrandom), all of them do contribute to the width of all confidence intervals associated with the measurement result itself, for every possible level of confidence. This means that the variable employed to represent the measurement result cannot be a generic fuzzy variable of type 2, but it must adhere, at least, to the condition that both membership functions are normal. Therefore, fuzzy variables of type 2 like the one in Fig. 4.3b must be discarded.

Moreover, let us consider that, when random uncertainty contributions are present, a known, or supposed, probability distribution function of the possible values is given. As stated, this probability distribution function must be converted into a possibility distribution function, which is to say, the membership function of a fuzzy variable must be derived from the probability distribution function. This fuzzy variable, as shown in Chapter 5, has always a single peak value. This means that the interval of confidence of type 2 corresponding to level $\alpha = 1$ always degenerates into an interval of confidence of type one. That is,

$$b_1^{\alpha=1} = b_2^{\alpha=1}$$

$$b_3^{\alpha=1} = b_4^{\alpha=1}$$

Therefore, condition (4.8) is always satisfied and fuzzy variables of type 2 like the one in Fig. 4.3a must be discarded.

This means that only fuzzy variables of type 2 like the one in Fig. 4.5 can suitably represent measurement results. This last kind of variables is called, here, for the sake of brevity, random–fuzzy variables (RFVs).

The reason for this name is that it summarizes all of the above considerations. Indeed, because random variables are suitable to represent measurement results affected by random contributions, and fuzzy variables are suitable to represent measurement results affected by systematic or unknown contributions, the natural name for the object that is aimed at representing measurement results affected by all kinds of uncertainty contributions appears to be random–fuzzy variable.

In accordance to the definitions given above and in the previous chapters, it can be stated that, given an RFV X, the intervals of confidence of type 2 X_α, at the different levels α, are the α-cuts of the RFV. Each α-cut is an interval of confidence, within which the measurement result is supposed to lie with a given level of confidence. Similar to fuzzy variables of type 1, the confidence level of each α-cut is given by the value assumed by the necessity function on that α-cut:

$$\mathrm{Nec}(X_\alpha) = 1 - \alpha^-$$

Moreover, an interval of confidence of type 2 provides additional useful information, that is, the distinction between an internal interval X_i^α and an external one X_e^α, where $X_e^\alpha \equiv X_\alpha$.

The external interval X_e^α considers, of course, all contributions to uncertainty affecting the measurement result. This means that, for each level α, X_e^α represents the interval of confidence within which the measurement result is supposed to lie, with its associated level of confidence. Moreover, the knowledge of the internal interval X_i^α also allows one to know which part of the confidence interval is due to unknown systematic contributions to uncertainty and total ignorance and which part is due to random contributions. In fact, for each level α, the width of the internal interval quantifies the uncertainty contributions due to unknown systematic effects and total ignorance. On the other hand, the total width of the side intervals (which are obtained as $X_e^\alpha - X_i^\alpha$) quantifies the random contributions to uncertainty.

It follows that the two membership functions, which are determined by the sequence X_α of all confidence intervals of type 2, may naturally separate the contributions to measurement uncertainty due to random phenomena from those due to total ignorance and unknown systematic phenomena. In particular, the external membership function considers the total contribution to uncertainty, due to all contributions affecting the measurement result, whereas the internal membership function considers all nonrandom contributions.

It can be noted that, if $x_1^\alpha = x_2^\alpha$ and $x_3^\alpha = x_4^\alpha$ for every α between 0 and 1, which means that the two membership functions coincide, the RFV degenerates into a simple fuzzy variable; if $x_2^\alpha = x_3^\alpha$, for every α between 0 and 1, the RFV has a nil internal membership function, that is, only random phenomena are present; if $x_1^\alpha = x_2^\alpha = x_3^\alpha = x_4^\alpha$, for every α between 0 and 1, the RFV degenerates into a crisp variable.

The above considerations show, in an intuitive way, that the random–fuzzy variables can be effectively employed to represent the result of a measurement together with its uncertainty, provided that several issues are defined and the membership functions are built starting from the available information, obtained either experimentally or from a priori knowledge. These issues are covered in the subsequent chapters.

5

Construction of Random–Fuzzy Variables

In Chapter 4, random–fuzzy variables have been defined and presented as a suitable solution for representing the result of a measurement together with its associated uncertainty, no matter what its origin. In fact, an RFV is represented by two membership functions, which are naturally capable of considering the different kinds of uncertainty contributions separately.

The two membership functions are possibility distribution functions.

The internal one considers only unknown systematic contributions and total ignorance, and it is built, on the basis of the available information, according to the possibility theory. On the other hand, the external membership function considers all contributions and, according to the considerations done in Chapter 4, must be built, starting from the internal membership function, on the basis of the pure random contributions to uncertainty. This operation is the most delicate to perform, because random contributions are generally given in terms of a probability distribution function $p(x)$, and it is therefore necessary to perform a transformation from a probability distribution function into a possibility distribution function $r(x)$ [MLF97].

The next section discusses some probability–possibility transformations available in the literature and shows that these transformations are not totally compatible with the concept of random–fuzzy variable proposed in Chapter 4. Therefore, Section 5.2 defines a specific probability–possibility transformation that fits with the random–fuzzy variables considered in this book.

5.1 Probability–possibility transformations

The probability–possibility transformations defined in the literature are bi-univocal functions $\mathbf{p} \leftrightarrow \mathbf{r}$, which allow one both to convert a probability distribution function into a possibility distribution function and to convert a possibility distribution function into a probability distribution function. However, for the aim of this book, only the univocal transformation $\mathbf{p} \rightarrow \mathbf{r}$ is needed. For this reason, in the following, only the univocal functions $\mathbf{p} \rightarrow \mathbf{r}$, or the $\mathbf{p} \rightarrow \mathbf{r}$ part of the biunivocal functions $\mathbf{p} \leftrightarrow \mathbf{r}$, will be considered and discussed.

In order to discuss these transformations, let us introduce the following conventional notations. Let $X = \{x_1, x_2, \ldots, x_n\}$ be the universal set, and let $p_i = p(x_i)$ and $r_i = r(x_i)$ be the probability and possibility of element x_i, respectively. Because of the normalization conditions defined in the probability and possibility theories, the following applies, in the discrete-time domain:

$$\max(r_i) = 1$$

and

$$\sum_{i=1}^{n} p(x_i) = 1$$

On the other hand, in the continuous-time domain, it is

$$\sup(r_i) = 1$$

and

$$\int_{\Re} p_i = 1$$

A $\mathbf{p} \rightarrow \mathbf{r}$ transformation is aimed at evaluating values r_i, starting from known values p_i.

The most common transformation is based on the *ratio scale*:

$$r_i = p_i \, c$$

for $i = 1, \ldots, n$, where c is a positive constant. In particular, in order to satisfy the above conditions, the following applies:

$$r_i = \frac{p_i}{\max(p_i)} \tag{5.1}$$

The above relationship is valid for both the discrete-time domain and the continuous-time domain, because it represents a simple normalization to one.

An example of this transformation is shown in Fig. 5.1. The solid line shows a Gaussian probability distribution function with unitary subtended area, mean value $\mu = 1$, and standard deviation $\sigma = 0.2$; the dashed line shows the corresponding possibility distribution function. From this figure and the above equation, it is possible to state that this $\mathbf{p} \rightarrow \mathbf{r}$ transformation always maintains the shape of the distribution function.

Another type of transformations is based on the *interval scale*:

$$r_i = p_i \, c + d$$

for $i = 1, \ldots, n$, where c and d are positive constants.

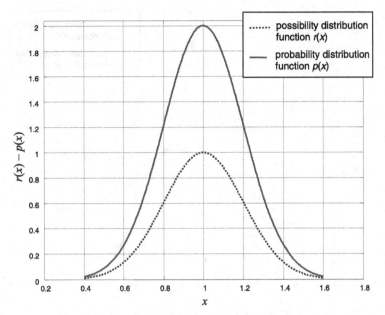

Fig. 5.1. p → r transformation according to Eq. (5.1).

An example of this kind of transformation is given by the following equation:

$$r_i = c \cdot (p_i - \max(p_i)) + 1 \qquad (5.2)$$

which still depends on the value assigned to constant c, and hence, it is not unique. Anyway, constant c cannot range from zero to infinite, but it is upper limited by the value:

$$c_{\max} = \frac{1}{\max(p_i)}$$

In fact, the application of Eq. (5.2) with values of c greater than $c_{\max}$ leads to negative values for r_i, and this is not allowed, by definition, for the possibility distribution function.

Moreover, when $c = c_{\max}$, Eqs. (5.1) and (5.2) become the same.

Equation (5.2) only applies in the discrete-time domain. Let us consider, however, that, in the continuous-time domain, the same formula can be used, after having properly scaled values p_i for a suitable positive constant.

Figure 5.2 shows some examples of application of Eq. (5.2) for different values of c. The initial probability distribution function is the same as that in Fig. 5.1.

It can be noted from Eq. (5.2) that, in the limit case when $c = 0$, the obtained possibility distribution function describes a rectangular fuzzy variable. When the value of c increases, the obtained possibility distribution functions assume the shapes shown in Fig. 5.2.

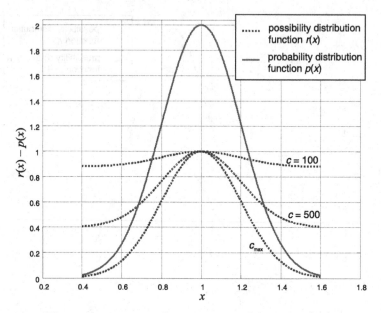

Fig. 5.2. $\mathbf{p} \rightarrow \mathbf{r}$ transformation according to Eq. (5.2).

Another kind of $\mathbf{p} \rightarrow \mathbf{r}$ transformation is described by the following equation:

$$r_i = \sum_{j=1}^{n} \min(p_i, p_j) \tag{5.3}$$

for $i = 1, \ldots, n$.

Also Eq. (5.3) applies only in the discrete-time domain. However, the same considerations done for Eq. (5.2) can be applied, and hence, the formula can be easily used also in the continuous-time domain.

Figure 5.3 shows an example of this transformation, where the initial probability distribution function is the same as that in Fig. 5.1.

5.2 A specific probability–possibility transformation for the construction of RFVs

The probability–possibility transformations described in the previous section are not suitable for the construction of random–fuzzy variables, because all of them are independent of the concepts of intervals of confidence and levels of confidence, which are the basis of RFVs. In fact, let us remember that an RFV can be seen as a set of confidence intervals, its α-cuts, at the different levels of confidence. Hence, the suitable way to transform the information contained

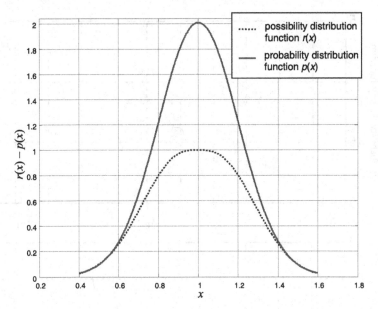

Fig. 5.3. p → r transformation according to Eq. (5.3).

in a probability distribution function into a possibility distribution function should somehow be based on the concept of intervals of confidence.

A suitable transformation can therefore be defined by considering that this concept is present in both the probability and the possibility theories. In particular, given a probability distribution function $p(x)$, the confidence interval associated with a particular level of confidence λ (where $0 \leq \lambda \leq 1$) is the interval $[x_a, x_b]$ for which

$$\int_{x_a}^{x_b} p(x) \, dx = \lambda \tag{5.4}$$

On the other hand, given a fuzzy, or a random–fuzzy variable X, the level of confidence associated with the generic α-cut, that is, with the generic confidence interval, is given by the necessity measure defined over the α-cut itself, such as

$$\mathrm{Nec}(X_\alpha) = 1 - \alpha^- \tag{5.5}$$

or

$$\mathrm{Nec}(X_\alpha) = 1 - \alpha \tag{5.6}$$

whenever the number of the considered levels α is high enough, as shown in Chapter 2.

The considered original transformation is based on Eqs. (5.4) and (5.6). In fact, if interval $[x_a, x_b]$ is known, together with its associated level of confidence

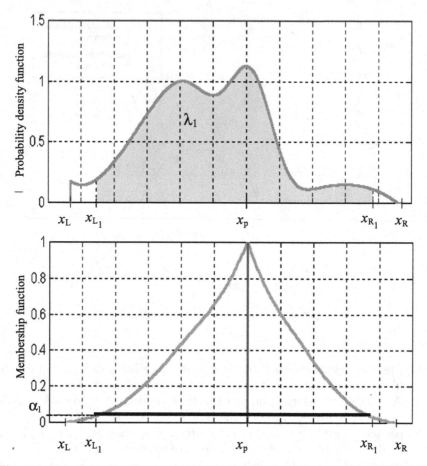

Fig. 5.4. Transformation from a probability density function into a possibility density function. Level α_1 is determined by $\alpha_1 = 1 - \lambda_1$.

λ, then, it is possible to state that $[x_a, x_b]$ is the α-cut of a fuzzy variable at level:

$$\alpha = 1 - \lambda \qquad (5.7)$$

Let us consider a unimodal[1] probability density function $p(x)$ defined over interval $[x_L, x_R]$, as shown in Fig. 5.4. Since, by definition of $p(x)$, it is

---

[1] This assumption implies that the bimodal probability distribution functions cannot be transformed into a possibility distribution function. From a strict mathematical point of view, this is perfectly coherent with the assumption that a possibility distribution function must be convex. However, bimodal probability distribution functions can be found in the measurement practice, when the result of a measurement is supposed to concentrate toward the edges of the interval within which the values of the measurand are supposed to lie. This case can be

$$\int_{x_L}^{x_R} p(x)\ dx = 1$$

interval $[x_L, x_R]$ represents the confidence interval to which a level of confidence equal to one is associated.

In case the probability distribution function is not limited, it is still possible to define an interval $[x_L, x_R]$. In fact, in this case, the probability distribution function decreases asymptotically to zero[2] and hence it is possible to neglect values outside a certain interval. For instance, in case $p(x)$ is a Gaussian function with standard deviation σ, the interval $[-3\sigma; +3\sigma]$ around the mean value could be chosen. This would mean to consider the interval of confidence at level of confidence 0.997, assign this interval to the confidence interval at level of confidence 1, and neglect all values outside it.

Once interval $[x_L, x_R]$ is determined, this same interval is considered the support of the possibility distribution function $r(x)$, corresponding to the given $p(x)$. Hence, according to Eq. (5.7), interval $[x_L, x_R]$ is the α-cut, at level $\alpha = 0$, of the membership function we want to determine.

In this way, a perfect agreement is obtained between the information given by $p(x)$ and $r(x)$ about the interval of confidence at level of confidence 1.

Let us now consider value x_p, for which $p(x)$ reaches its absolute maximum. Then, the peak value of the desired membership function is assumed to fall at this same value. This means that the degenerated interval $[x_p, x_p]$ corresponds to the α-cut for $\alpha = 1$ of the membership function. Again a perfect agreement is obtained between the information given by $p(x)$ and $r(x)$ about this interval of confidence. In fact, when $r(x)$ is considered, the necessity function associated with α-cut $[x_p, x_p]$ is zero, and when $p(x)$ is considered, the level of confidence associated with $[x_p, x_p]$ is

$$\int_{x_p}^{x_p} p(x)\ dx = 0$$

In case the probability density function presents two absolute maximum values, the value x_p is associated with the mean value of the two ones, by definition.

At this point, the α-cuts for $\alpha = 0$ and $\alpha = 1$ have been found, whereas the intermediate α-cuts must still be considered. To do so, let us remember that the α-cuts of a fuzzy variable can be ordered in a set of nested intervals.

In this respect, let us consider interval $[x_L, x_p]$, already defined. Let us take, within this interval, M points $x_{L_1}, x_{L_2}, \ldots, x_{L_M}$, where $x_L < x_{L_1} < x_{L_2} < \cdots < x_{L_M} < x_p$. In this way, interval $[x_L, x_p]$ is divided into $M+1$ subintervals

---

seen as if two different measurement results are possible, and therefore, these two results can be treated as two different RFVs.

[2] Let us consider that this requirement is satisfied by all not limited probability distribution function. In fact, only under this assumption, integral $\int_{x_L}^{x_R} p(x)\ dx$ assumes a finite value.

having the same width: $[x_L, x_{L_1}], [x_{L_1}, x_{L_2}], \ldots, [x_{L_k}, x_{L_{k+1}}], \ldots, [x_{L_{M-1}}, x_{L_M}],$ $[x_{L_M}, x_p]$. The number of points M is chosen according to the required resolution. Of course, the greater is M, the greater is the resolution of the desired fuzzy variable.

Let us now consider interval $[x_p, x_R]$, already defined, and take M points $x_{R_1}, x_{R_2}, \ldots, x_{R_M}$, where $x_p < x_{R_M} < x_{R_{M-1}} < \cdots < x_{R_1} < x_R$, so that $M + 1$ subintervals with the same width are again determined: $[x_p, x_{R_M}],$ $[x_{R_M}, x_{R_{M-1}}], \ldots, [x_{R_2}, x_{R_1}], [x_{R_1}, x_p]$.

The following $M + 2$ intervals can be now taken into account: $[x_p, x_p],$ $[x_{L_M}, x_{R_M}], \ldots, [x_{L_k}, x_{R_k}], \ldots, [x_{L_1}, x_{R_1}], [x_L, x_R]$.

In this sequence, it can be noted that each interval is included in the following one and includes the previous one. Therefore, a set of nested intervals is obtained. Hence, these intervals can be the α-cuts of a fuzzy variable. Therefore, these intervals are taken as the α-cuts of the desired fuzzy variable, for specific levels α. The correct values α can be associated with intervals $[x_p, x_p], [x_{L_M}, x_{R_M}], \ldots, [x_{L_k}, x_{R_k}], \ldots, [x_{L_1}, x_{R_1}], [x_L, x_R]$ starting from the probability distribution function $p(x)$, according to Eq. (5.7).

Levels zero and one have been already associated with intervals $[x_L, x_R]$ and $[x_p, x_p]$, respectively. Then, for $1 \leq k \leq M$, it is necessary to associate a level α with each α-cut $[x_{L_k}, x_{R_k}]$.

It is known that, if the probability distribution function $p(x)$ is considered, the level of confidence associated with the generic interval $[x_{L_k}, x_{R_k}]$ is given by

$$\int_{L_k}^{R_k} p(x) \, dx = \lambda_k$$

Therefore, the level:

$$\alpha_k = 1 - \lambda_k$$

is associated with α-cut $[x_{L_k}, x_{R_k}]$.

Figure 5.4 shows an example of this construction for interval $[x_{L_1}, x_{R_1}]$.

If the same Gaussian probability distribution function as that in the example of Fig. 5.1 is considered, the membership function reported in Fig. 5.5 is obtained. This figure shows that this $\mathbf{p} \rightarrow \mathbf{r}$ transformation does not maintain the shape of the probability distribution function, as the transformations shown in Figs. 5.1, 5.2, and 5.3 do. However, this was not an essential requirement; on the contrary, the important and essential requirement that the initial probability density function and the final possibility density function contain the same information from a metrological point of view is met. In fact, both of them provide, for each desired level of confidence, the same interval of confidence. Moreover, it can be also stated that the description in terms of a fuzzy variable is simpler. In fact, although the description in terms of a probability distribution function requires, in order to obtain the confidence interval for a given confidence level, to perform an integral, the description in terms of a fuzzy variable is more immediate, because confidence intervals are exactly the α-cuts of the fuzzy variable.

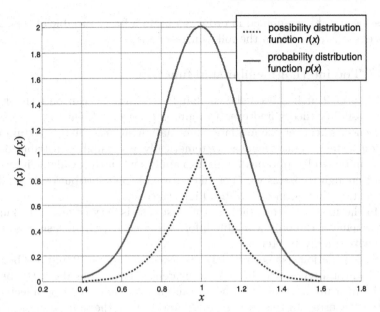

Fig. 5.5. p → r transformation according to Eq. (5.7).

The next section shows this transformation applied to the construction of an RFV.

5.3 Construction of RFVs

According to the definition given in Chapter 4, a random–fuzzy variable is a special case of fuzzy variable of type 2 and is defined by two membership functions. As shown in that chapter, it can be stated that the internal membership function considers all kinds of unknown contributions to measurement uncertainty, including the unknown systematic ones, whereas the external membership function also considers random contributions. Random contributions are therefore taken into account in that part of the RFV between the two membership functions.

In other words, if the different intervals of confidence of type 2 of the RFV are considered, it can be stated that, for each considered interval, the internal one considers systematic and unknown contributions, whereas the external ones consider random contributions.

Because of the different meaning of the internal and the external intervals defined by the interval of confidence of type 2, different methods are used to determine them. Both methods allow one to build a fuzzy variable of type 1. Therefore, whenever both random and nonrandom contributions to uncertainty affect the measurement result, the two type 1 fuzzy variables must be suitably combined in order to obtain a fuzzy variable of type 2. This procedure is reported in the following sections.

5.3.1 The internal membership function

In Chapter 2, it has been shown that fuzzy variables can be framed within the possibility theory and used to represent degrees of belief.

Hence, whenever the available information suggests that a certain contribution to uncertainty behaves systematically, a possibility distribution function can be readily obtained. This also applies when the available information is not enough to assess whether the nature of the contribution is random or systematic, that is, in case of total ignorance.

In the measurement practice, this last case is very common. In fact, it is met whenever the only available information is that the measurement result falls within a given interval.

This situation has been widely discussed in Chapter 2 and has been classified as total ignorance. In fact, no evidence is available about the behavior of the measurement result. The measurement result could be affected by a systematic error. In this case, it would always take the same value, within the given interval. On the other hand, the measurement result could be affected by a random error. In this second case, it could probably take all different values of the given interval, according to a certain (unknown[3]) probability distribution function.

As shown in Chapter 2, this situation can be fully described only by a fuzzy variable with a rectangular membership function.

Another very common situation in the measurement practice is the presence of a systematic error. In this case, different situations can be considered, according to the available information.

The most common one is very similar to the one described above: The measurement result, because of the effect of the systematic contribution, surely falls within a given interval. In this case, if no other information is available, the measurement result is represented again by a rectangular membership function, as shown in Chapter 2.

On the other hand, if it is known, for instance, that the measurement algorithm compensate for the considered systematic effect, the contribution itself must not be considered.

Another possible situation, although not very common, is that the available information suggests different degrees of beliefs for different intervals. Also this situation has been discussed in Chapter 2, where it has been shown that, in this case, it is possible to associate it with a trapezoidal membership function, a triangular membership function, or in general, a membership function of whichever shape.

---

[3] As widely discussed in Chapter 2, the rectangular probability distribution, generally used by statisticians to represent total ignorance, indeed is not suitable to represent total ignorance.

Hence, the construction of the internal membership function of a random–fuzzy variable is quite immediate. In fact, the available information is always already provided in terms of degrees of belief, which readily correspond to possibility distribution functions.

5.3.2 The external membership function

As discussed at the beginning of this chapter, the shape of the external membership function is determined by the probability distribution function, which describes the distribution of the possible measurement results.

This probability distribution function is always known. In fact, it can be either determined in terms of a relative frequency histogram, when experimental data are available (type A evaluation), or assumed starting from the available information (type B evaluation). In the first case, if a sufficient number of measurement results is available, the histogram can be suitably interpolated in order to approximate the probability distribution function $p(x)$ of the results themselves.

Function $p(x)$ must be then transformed into a possibility distribution function $r(x)$, by applying the probability–possibility transformation described in Section 5.2.

5.3.3 Combination of the two membership functions

When no random contributions affect the measurement result, this last one can be represented by an RFV, for which the external membership function coincide with the internal one, built as shown in Section 5.3.1.

When only random contributions affect the measurement result, this last one can be still represented by an RFV, for which the internal membership function is nil (this means that the width of all its α-cuts is zero) and the external one is built as shown in Section 5.3.2.

In case the available information shows the presence of both systematic, and/or unknown, and random contributions, the associated random–fuzzy variable is obtained by suitably combining the two above methods.

Let us suppose that the application of those methods leads to two membership functions μ_1 and μ_2, respectively. In a very intuitive way, it can be stated that the combination of μ_1 and μ_2 is obtained by "inserting" the membership function μ_1 into the membership function μ_2. In this way, the final random–fuzzy variable will have an internal membership function, which takes into account all systematic and unknown contributions, and an external membership function, which takes into account all kinds of contributions. The internal membership function is μ_1, whereas the external membership function is built taking into account both μ_1 and μ_2.

Let us consider this combination from a mathematical point of view. Let us consider the generic α-cuts, at the same level α, of the two membership

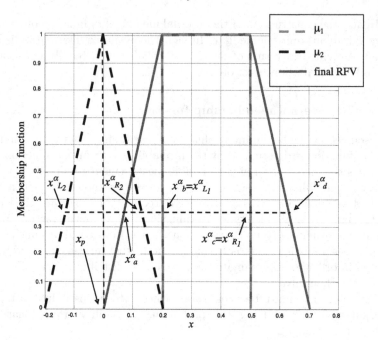

Fig. 5.6. Combination of membership functions μ_1 and μ_2.

functions μ_1 and μ_2, and let us denote them $[x_{L_1}^\alpha, x_{R_1}^\alpha]$ and $[x_{L_2}^\alpha, x_{R_2}^\alpha]$, respectively.

According to the indications given in the previous section, interval $[x_{L_2}^\alpha, x_{R_2}^\alpha]$ can be divided into the two intervals $[x_{L_2}^\alpha, x_p]$ and $[x_p, x_{R_2}^\alpha]$, where x_p corresponds to the peak value of μ_2, as also shown in Fig. 5.4.

The α-cut $X_\alpha = \{x_a^\alpha, x_b^\alpha, x_c^\alpha, x_d^\alpha\}$ of the final random–fuzzy variable can be defined as

$$x_b^\alpha = x_{L_1}^\alpha$$

$$x_c^\alpha = x_{R_1}^\alpha$$

$$x_a^\alpha = x_b^\alpha - \left(x_p - x_{L_2}^\alpha\right)$$

$$x_d^\alpha = x_c^\alpha + \left(x_{R_2}^\alpha - x_p\right)$$

If these equations are applied to all α-cuts, the final random–fuzzy variable is obtained. An example is shown in Fig. 5.6.

6

Fuzzy Operators

Until now, the previous chapters have shown how uncertainty in measurement can be represented within the Theory of Evidence and have shown how the result of a measurement can be suitably expressed, together with its uncertainty, in terms of a random–fuzzy variable (RFV).

In order to fully assess the capability of RFVs to represent uncertainty in measurement, it is necessary to define suitable mathematical operators, so that the different contribution to uncertainty can be combined together when indirect measurements are considered.

Therefore, this chapter is aimed at introducing some basic fuzzy operators, which will be extensively applied in the next chapters to define the mathematics of the RFVs. The operators considered in this chapter are defined on fuzzy variables, but all given definitions can be readily extended to the random–fuzzy variables, by operating on both of their membership functions.[1]

6.1 Aggregation operations

The aggregation operations on fuzzy variables are operations by which several fuzzy variables are combined in order to produce a single fuzzy variable.

Formally, any aggregation operation on n fuzzy variables $A_1, A_2, \ldots, A_n$ (with $n \geq 2$), defined on the same universal set X, is defined by a function:

$$h : [0,1]^n \to [0,1]$$

For each element x of the universal set, the previous function h takes, as its arguments, the membership grades associated with x by the fuzzy sets

---

[1] This is the most general case. However, in some applications, it could be enough to consider the external membership functions only, because they contain the whole information. This simplification generally applies when RFVs are employed to represent measurement results.

$A_1, A_2, \ldots, A_n$. In other words, function h produces an *aggregate fuzzy variable* A by suitably combining the membership grades:

$$A(x) = h(A_1(x), A_2(x), \ldots, A_n(x))$$

for all $x \in X$.

The previous equation shows that function h is independent from x and depends only on values $A_1(x), A_2(x), \ldots, A_n(x)$. Therefore, in the following, the arguments of the aggregation operation h, $A_1(x), A_2(x), \ldots, A_n(x)$, will be considered arbitrary numbers $a_1, a_2, \ldots, a_n \in [0, 1]$.

The following definition defines the properties that must be satisfied by any aggregation operators.

Definition. *A mapping $h : [0,1]^n \to [0,1]$ is an aggregation function if and only if it satisfies the following three axiomatic requirements:*

- *Boundary condition*

$$h(0, 0, \ldots, 0) = 0 \quad and \quad h(1, 1, \ldots, 1) = 1 \tag{6.1}$$

- *Continuity*
$$h \text{ is a continuous function} \tag{6.2}$$

- *Monotonicity*
  For any pair $\langle a_1, a_2, \ldots, a_n \rangle$ and $\langle b_1, b_2, \ldots, b_n \rangle$ of n-tuples such that $a_i, b_i \in [0,1]$ for all $i \in N_n$,

$$h(a_1, a_2, \ldots, a_n) \leq h(b_1, b_2, \ldots, b_n) \quad if \quad a_i \leq b_i \quad for\ all\ i \in N_n \tag{6.3}$$

Equation (6.3) shows that an aggregation operation is monotonically increasing in all its arguments.

Besides these essential requirements, aggregation operations on fuzzy variables are usually expected to satisfy the other two additional axiomatic requirements. The first one is the requirement of simmetricity in all its arguments:

- *Simmetricity*

$$h(a_1, a_2, \ldots, a_n) = h(a_{p(1)}, a_{p(2)}, \ldots, a_{p(n)}) \tag{6.4}$$

for any permutation p on N_n.

The second one is one among the following:

- *Idempotency*

$$h(a, a, \ldots, a) = a \quad for\ all\ a \in [0, 1] \tag{6.5}$$

- *Subidempotency*

$$h(a, a, \ldots, a) \leq a \quad \text{for all } a \in [0, 1] \qquad (6.6)$$

- *Superidempotency*

$$h(a, a, \ldots, a) \geq a \quad \text{for all } a \in [0, 1] \qquad (6.7)$$

Axiom (6.4) reflects the usual assumption that the aggregated fuzzy variables are equally important. Axioms (6.5) states that any aggregation of equal fuzzy variables should result in the same fuzzy variable. On the other hand, the alternative axioms (6.6) and (6.7) are weaker requirements. In fact, when the membership grades of the initial fuzzy variables are equal, let us say a, they allow that the resulting membership grade is, respectively, lower or greater than this same value a. The three requirements (6.5), (6.6), and (6.7) define, respectively, three different subclasses of the aggregation operators: the averaging operators, the t-norm, and the t-conorm.

6.1.1 t-norm

The term 't-norm' stands for triangular norm. Triangular norms were introduced for the first time to model the distances in probabilistic metric spaces [SS63]. In fuzzy set theory, the triangular norms are extensively used to model the logical connective *and* or, in other words, the intersection of two fuzzy sets. For this reason, the terms 't-norms' and 'fuzzy intersections' can be used indifferently.

Let us now consider $n = 2$. Then, the t-norm of two fuzzy variables A and B is a mapping:

$$T : [0, 1] \times [0, 1] \rightarrow [0, 1] \qquad (6.8)$$

For each element x of the universal set, function (6.8) takes, as its arguments, the membership grades associated with x by the fuzzy sets A and B, respectively:

$$(A \cap B)(x) = T[A(x), B(x)] \qquad (6.9)$$

Equation (6.9) shows that function T is independent from x and depends only on values $A(x)$ and $B(x)$. Therefore, in the following, the arguments of the t-norm, $A(x)$ and $B(x)$, will be considered arbitrary numbers $a, b, \in [0, 1]$.

Let us now consider the following definition of t-norm, which describes its properties.

Definition. *A mapping* $T : [0, 1] \times [0, 1] \rightarrow [0, 1]$ *is a triangular norm if and only if it satisfies the following properties for all* $a, b, c, d \in [0, 1]$:

- *Symmetricity*

$$T(a, b) = T(b, a) \qquad (6.10)$$

- *Associativity*

$$T(a, T(b, d)) = T(T(a, b), d) \tag{6.11}$$

- *Monotonicity*

$$T(a, b) \leq T(c, d) \tag{6.12}$$

$$if\ a \leq c\ and\ b \leq d$$

- *One identity*

$$T(a, 1) = a \tag{6.13}$$

The first three properties require that the t-norm is symmetric (or commutative), associative, and nondecreasing in each argument, whereas the last property defines the boundary condition.

Axioms (6.10), (6.12), and (6.13) ensure that the fuzzy intersection defined by (6.9) becomes the classic set intersection when sets A and B are crisp.

In fact, from the boundary condition, it follows that

$$T(0, 1) = 0 \quad \text{and} \quad T(1, 1) = 1$$

Then, from simmetricity and monotonicity, it follows respectively that

$$T(1, 0) = 0 \quad \text{and} \quad T(0, 0) = 0$$

Moreover, simmetricity ensures that the fuzzy intersection is indifferent to the order with which the sets to be combined are considered.

The boundary and simmetricity conditions ensure that, as it can be also intuitively stated, when one argument of T is one, the membership grade of the intersection is equal to the other argument.

It can be also intuitively stated that a decrement in the degree of membership in A or B cannot produce an increment in the degree of membership of the intersection. This is ensured, in a strict way, by monotonicity and simmetricity.

At last, axiom (6.11) allows one to extend the operation of fuzzy intersection to more than two sets. In fact, associativity ensures that the intersection of any number of sets is independent on the order with which sets are pairwise grouped.

Let us note that, thanks to their property of associativity, the definition of fuzzy t-norm, which has been given only for two arguments, can be easily extended to any number of arguments.

The class of the triangular norms can be restricted by considering various additional requirements. Three of the most important requirements are expressed by the following axioms:

- *Continuity*

$$T \quad \text{is a continuous function} \tag{6.14}$$

- *Subidempotency*

$$T(a,a) \leq a \qquad (6.15)$$

- *Strict monotonicity*

$$T(a,b) < T(c,d) \qquad (6.16)$$

$$\text{if } a < c \text{ and } b < d$$

Axiom (6.14) prevents situations in which small changes in the membership grade of either A or B would produce large changes in the membership grade of the intersection.

Axiom (6.15) requires that the membership grade of the intersection, whenever, for a given value x, both the membership grades in A and B assume the same value, say a, must not exceed this same value a. As this requirement is weaker than $T(a,a) = a$, which, as already stated, is called *idempotency*, it is called *subidempotency*.

Finally, axiom (6.16) expresses a stronger form of monotonicity.

Different t-norms can be defined. For example, the standard fuzzy intersection is defined as[2]

$$T(a,b) = \min(a,b) = \min\{a,b\}$$

for all $a, b, \in [0,1]$.

The following theorem states that this t-norm is the only one that satisfies $T(a,a) = a$, that is, the equality in relationship (6.15).

Theorem 6.1. *The standard fuzzy intersection is the only idempotent t-norm.*

Proof:

Clearly, $\min\{a,a\} = a$ for all $a \in [0,1]$. Assume that a t-norm exists such that $T(a,a) = a$ for all $a \in [0,1]$. Then, for any $a,b \in [0,1]$, if $a \leq b$, then

$$a = T(a,a) \leq T(a,b) \leq T(a,1) = a$$

by monotonicity and the boundary condition. Hence,

$$T(a,b) = a = \min\{a,b\}$$

Similarly, if $a \geq b$, then

$$b = T(b,b) \leq T(a,b) \leq T(1,b) = b$$

and, consequently, $T(a,b) = b = \min\{a,b\}$. Hence, $T(a,b) = \min\{a,b\}$ for all $a,b \in [0,1]$.

---

[2] Let us consider the symbols $\min\{a,b\}$ and $\min(a,b)$. In the following, the first one denotes the minimum among the values in the braces, whereas the second one denotes the standard fuzzy intersection among the aggregates in the round brackets.

A continuous t-norm that satisfies subidempotency is called the *Archimedean t-norm*; if it also satisfies strict monotonicity, it is called the *strict Archimedean t-norm*.

Different methods are available to obtain t-norms and to combine given t-norms in order to obtain new t-norms. However, this subject is not interesting for the aim of this book; a few significant examples of some t-norms that are frequently used as fuzzy intersection are given here, and the reader is referred to the literature [KY95], [SS63] for a more comprehensive discussion.

For all $a, b \in [0, 1]$, the following t-norms are defined:

- Lukasiewicz:
$$T_L(a, b) = \max\{a + b - 1, 0\}$$

- product:
$$T_P(a, b) = a\, b$$

- weak:
$$T_W(a, b) = \begin{cases} \min\{a, b\} & \text{if } \max\{a, b\} = 1 \\ 0 & \text{otherwise} \end{cases}$$

- Hamacher: for $\gamma \geq 0$,
$$H_\gamma(a, b) = \begin{cases} T_W(a, b) & \text{if } \gamma = \infty \\ \dfrac{a\, b}{\gamma + (1 - \gamma)\,(a + b - ab)} & \text{otherwise} \end{cases}$$

- Dubois and Prade: for $\alpha \in [0, 1]$,
$$D_\alpha(a, b) = \begin{cases} \min(a, b) & \text{if } \alpha = 0 \\ T_P(a, b) & \text{if } \alpha = 1 \\ \dfrac{a\, b}{\max\{a, b, \alpha\}} & \text{if } \alpha \in (0, 1) \end{cases}$$

- Yager: for $p \geq 0$,
$$Y_p(a, b) = \begin{cases} T_W(a, b) & \text{if } p = 0 \\ T_L(a, b) & \text{if } p = 1 \\ 1 - \min\{1, \sqrt[p]{[(1 - a)^p + (1 - b)^p]}\} & \text{otherwise} \end{cases}$$

- Frank: for $\lambda \geq 0$,
$$F_\lambda(a, b) = \begin{cases} \min(a, b) & \text{if } \lambda = 0 \\ T_P(a, b) & \text{if } \lambda = 1 \\ T_L(a, b) & \text{if } \lambda = \infty \\ 1 - \log_\lambda \left[1 + \dfrac{(\lambda^a - 1)\,(\lambda^b - 1)}{\lambda - 1}\right] & \text{otherwise} \end{cases}$$

- Schweizer and Sklar:

$$SS_p(a,b) = \begin{cases} T_P(a,b) & \text{if } p = 0 \\ T_L(a,b) & \text{if } p = 1 \\ \dfrac{a\,b}{a+b-ab} & \text{if } p = -1 \\ T_W(a,b) & \text{if } p = +\infty \\ \min(a,b) & \text{if } p = -\infty \\ \sqrt[p]{\max\{0, a^p + b^p - 1\}} & \text{otherwise} \end{cases}$$

Figure 6.1 shows, for two generic fuzzy variables A and B, represented with the star-lines, some of the above defined fuzzy intersections: the standard fuzzy intersection, the Lukasiewicz intersection, the weak intersection, and the product intersection. This figure clearly shows that, for all $a, b \in [0,1]$,

$$T_W \leq T_L \leq T_P \leq \min(a,b)$$

Moreover, in this particular example, in which $\mu_B(x) \leq \mu_A(x)$, for every $x \in X$ (see Fig. 6.1), the standard intersection coincides with $\mu_B(x)$; that is, $\min(a,b)$ coincides with the variable with the smallest membership function.

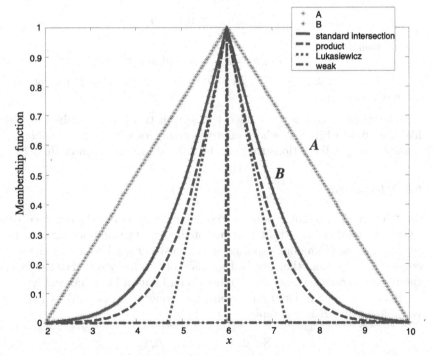

Fig. 6.1. Examples of different t-norms applied to fuzzy sets A and B.

The above inequality can be extended to every kind of fuzzy intersection. In fact, the following theorem states that every t-norm provides a fuzzy variable whose membership function falls between the membership function provided by the weak intersection and that provided by the standard intersection.

Theorem 6.2. *Given a fuzzy intersection* $T(a, b)$, *it is:*

$$T_W(a, b) \leq T(a, b) \leq \min(a, b)$$

for all $a, b, \in [0, 1]$

Proof:

Upper bound. By the boundary condition and monotonicity,

$$T(a, b) \leq T(a, 1) = a$$

and, by commutativity,

$$T(a, b) = T(b, a) \leq T(b, 1) = b$$

Hence, $T(a, b) \leq a$ and $T(a, b) \leq b$; that is, $T(a, b) \leq \min(a, b)$.

Lower bound. From the boundary condition $T(a, b) = a$ when $b = 1$ and $T(a, b) = b$ when $a = 1$. As $T(a, b) \leq \min(a, b)$ and $T(a, b) \in [0, 1]$, clearly,

$$T(a, 0) = T(0, b) = 0$$

By monotonicity,

$$T(a, b) \geq T(a, 0) = T(0, b) = 0$$

Hence, the weak intersection $T_W(a, b)$ is the lower bound of $T(a, b)$ for any $a, b \in [0, 1]$.

Therefore, it can be concluded that, given two fuzzy variables A and B like the ones in Fig. 6.1, where $\min(a, b)$ coincides with fuzzy variable B, all possible fuzzy intersections that can be defined are surely lower than B.

6.1.2 t-conorm

Together with triangular norms, triangular conorms are defined too, which represent another important subclass of the aggregation operators. In fuzzy set theory, the triangular conorms are extensively used to model the logical connective *or* or, in other words, the union of two fuzzy sets. For this reason, the terms 't-conorms' and 'fuzzy unions' can be used interchangeably.

If $n = 2$ is considered, the triangular conorm (t-conorm) of two fuzzy variables A and B is a mapping:

$$S : [0, 1] \times [0, 1] \rightarrow [0, 1] \tag{6.17}$$

For each element x of the universal set, function (6.17) takes, as its arguments, the membership grades associated with x by the fuzzy sets A and B, respectively:

$$(A \cup B)(x) = S[A(x), B(x)] \tag{6.18}$$

Equation (6.18) shows that function S is independent from x, and depends only on values $A(x)$ and $B(x)$. Therefore, in the following, the arguments of the t-conorm, $A(x)$ and $B(x)$, will be considered arbitrary numbers $a, b, \in [0,1]$.

Let us now consider the following definition of t-conorm, which describes its properties.

Definition. *A mapping $S : [0,1] \times [0,1] \to [0,1]$ is a triangular conorm if and only if it satisfies the following properties for all $a, b, c, d \in [0,1]$:*

- *Symmetricity*

$$S(a,b) = S(b,a) \tag{6.19}$$

- *Associativity*

$$S(a, S(b,d)) = S(S(a,b), d) \tag{6.20}$$

- *Monotonicity*

$$S(a,b) \le S(c,d) \tag{6.21}$$

$$\text{if } a \le c \text{ and } b \le d$$

- *Zero identity*

$$S(a,0) = a \tag{6.22}$$

The first three properties require that the t-conorm is symmetric (or commutative), associative, and nondecreasing in each argument, whereas the last property defines the boundary condition. If axioms (6.19)–(6.22) are compared with (6.10)–(6.13), it is clear that they differ only in the boundary condition. Hence, the axioms are justified on the same ground as those for fuzzy intersections.

Similar to t-norms, the class of t-conorms can also be restricted by considering some additional requirements. The most important ones are expressed by the following axioms:

- *Continuity*

$$S \text{ is a continous function} \tag{6.23}$$

- *Superidempotency*

$$S(a,a) \ge a \tag{6.24}$$

- *Strict monotonicity*

$$S(a,b) < S(c,d) \tag{6.25}$$

$$\text{if } a < c \text{ and } b < d$$

These axioms are similar to axioms (6.14)–(6.16) for fuzzy intersections, except for the requirement of subidempotency for fuzzy intersection, which is here replaced with that of superidempotency.

Let us note that, thanks to their property of associativity, the definition of fuzzy t-conorm, which has been given only for two arguments, can be easily extended to any number of arguments.

Different t-conorms can be defined, which satisfy axioms from (6.19) to (6.22). For example, the standard fuzzy union is defined as[3]

$$S(a,b) = \max(a,b) = \max\{a,b\}$$

for all $a, b \in [0,1]$.

The following theorem states that this t-conorm is the only one that satisfies $S(a,a) = a$, that is, the equality in relationship (6.24).

Theorem 6.3. *The standard fuzzy union is the only idempotent t-conorm.*

Proof:

The proof of this theorem is practically the same as the proof of Theorem 6.1 and is left to the reader.

A continuous t-conorm that satisfies superidempotency is called the *Archimedean t-conorm*; if it also satisfies strict monotonicity, it is called the *strict Archimedean t-conorm*.

It can be proved that, if T is a t-norm, then the following equation:

$$S(a,b) = 1 - T(1 - a, 1 - b) \tag{6.26}$$

defines a t-conorm.

When Eq. (6.26) is applied, we say that the triangular conorm S is derived from the triangular norm T. Let us prove, for instance, that the standard fuzzy union is derived from the standard fuzzy intersection.

Proof:

The standard fuzzy intersection is defined by $T(a,b) = \min\{a,b\}$. Hence, it is: $T(1 - a, 1 - b) = \min\{1 - a, 1 - b\}$. Moreover,

$$\min\{1 - a, 1 - b\} = \begin{cases} 1 - a & \text{if } a > b \\ 1 - b & \text{if } a < b \end{cases}$$

Then,

$$1 - T(1 - a, 1 - b) = \begin{cases} a & \text{if } a > b \\ b & \text{if } a < b \end{cases}$$

Finally:

$$1 - T(1 - a, 1 - b) = \max\{a,b\} = S(a,b)$$

where $S(a,b)$ is the standard fuzzy union.

---

[3] Let us consider the symbols $\max\{a,b\}$ and $\max(a,b)$. In the following, the first one denotes the maximum among the values in the braces, whereas the second one denotes the standard fuzzy union among the aggregates in the round brackets.

Different methods are available to obtain t-conorms and combine given t-conorms in order to obtain new t-conorms [KY95], [SS63]. The following examples are of some t-conorms that are frequently used as fuzzy unions.

- Lukasiewicz union:

$$S_L(a,b) = \min\{a + b - 1, 0\}$$

- probabilistic union:

$$S_P(a,b) = a + b - ab$$

- strong union:

$$S_S(a,b) = \begin{cases} \max\{a,b\} & \text{if } \min\{a,b\} = 0 \\ 1 & \text{otherwise} \end{cases}$$

- Hamacher union:

$$\text{HOR}_\gamma(a,b) = \frac{a + b - (2 - \gamma)ab}{1 - (1 - \gamma)ab}, \quad \gamma \geq 0$$

- Dubois and Prade union: for $\alpha \in [0,1]$,

$$\text{DOR}_\alpha(a,b) = \begin{cases} \max(a,b) & \text{if } \alpha = 0 \\ S_P(a,b) & \text{if } \alpha = 1 \\ 1 - \dfrac{(1 - a)(1 - b)}{\max\{(1 - a), (1 - b), \alpha\}} & \text{if } \alpha \in (0,1) \end{cases}$$

- Yager union:

$$\text{YOR}_p(a,b) = \min\{1, \sqrt[p]{a^p + b^p}\}, \quad p > 0$$

- Frank: for $\lambda \geq 0$,

$$\text{FOR}_\lambda(a,b) = \begin{cases} \max(a,b) & \text{if } \lambda = 0 \\ S_P(a,b) & \text{if } \lambda = 1 \\ S_L(a,b) & \text{if } \lambda = \infty \\ 1 - \log_\lambda\left[1 + \dfrac{(\lambda^{1-a} - 1)\,(\lambda^{1-b} - 1)}{\lambda - 1}\right] & \text{otherwise} \end{cases}$$

- Schweizer and Sklar:

$$\text{SSOR}_p(a,b) = \begin{cases} S_P(a,b) & \text{if } p = 0 \\ S_L(a,b) & \text{if } p = 1 \\ \dfrac{a + b - 2ab}{1 - ab} & \text{if } p = -1 \\ S_s(a,b) & \text{if } p = +\infty \\ \max(a,b) & \text{if } p = -\infty \\ 1 - \sqrt[p]{\max\{0, (1 - a)^p + (1 - b)^p - 1\}} & \text{otherwise} \end{cases}$$

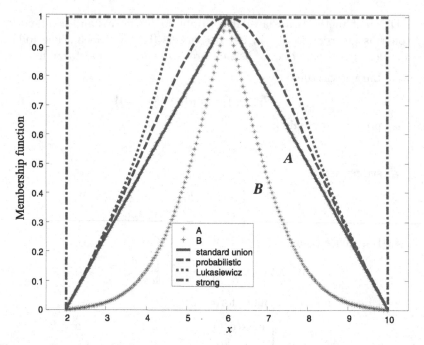

Fig. 6.2. Examples of different t-conorms applied to the same fuzzy sets A and B of the example of Fig. 6.1.

Figure 6.2 shows, for the same fuzzy variables A and B as those in Fig. 6.1, some of the above defined fuzzy unions: the standard fuzzy union, the Lukasiewicz union, the strong union, and the probabilistic union. This figure clearly shows that, for all $a, b \in [0, 1]$,

$$\max(a, b) \leq S_P \leq S_L \leq S_S$$

Moreover, in this particular example, in which $\mu_B(x) \leq \mu_A(x)$, for all $x \in X$ (see Fig. 6.2), the standard union coincides with $\mu_A(B)$; that is, $\max(a, b)$ coincides with the variable with the greatest membership function.

The above inequality can be extended to every kind of fuzzy union. In fact, the following theorem states that every t-conorm provides a fuzzy variable whose membership function falls between the membership function provided by the standard union and that provided by the strong union.

Theorem 6.4. *Given a fuzzy union $S(a, b)$, it is*

$$\max(a, b) \leq S(a, b) \leq S_s(a, b)$$

for all $a, b \in [0, 1]$

Proof:

The proof of this theorem is practically the same as the proof of Theorem 6.2, and is left to the reader.

Therefore, it can be concluded that, given two fuzzy variables A and B like the ones in Figs. 6.1 and 6.2, for which $\max(a, b)$ coincides with fuzzy variable A, all possible fuzzy unions are surely greater than A.

6.1.3 Averaging operations

As shown in the previous sections, the standard fuzzy intersection produces, for any given couple of fuzzy variables, the largest fuzzy variable among those produced by all possible t-norms. On the contrary, the standard fuzzy union produces the smallest fuzzy variable among the fuzzy variables produced by all possible t-conorms.

Let us consider two fuzzy variables A and B like the ones in Figs. 6.1 and 6.2, where $\mu_B(x) \leq \mu_A(x)$, for all $x \in X$. The standard fuzzy intersection between A and B provides, as a result, the fuzzy variable with the smallest membership function, that is, B; on the other hand, the standard fuzzy union between the same variables provides the fuzzy variable with the greatest membership function, that is, A. If other types of fuzzy intersections and fuzzy unions are considered, they provide fuzzy variables whose membership functions are, respectively, smaller than B and greater than A.

It can be noted that no t-norm or t-conorm ever 'fills' that part between the two fuzzy variables. An interesting question is whether there is some fuzzy operator able to do that. In fact, fuzzy intersections and fuzzy unions do not cover all operations by which fuzzy variables can be aggregated, but they are only particular subclasses of the aggregation operations. Another particular subclass is represented by the averaging operations.

Let us consider $n = 2$, for the sake of simplicity. The following definition sets the properties of an averaging operator.

Definition. *A mapping $M : [0, 1] \times [0, 1] \to [0, 1]$ is an averaging operation if and only if it satisfies the following properties for all $a, b, c, d \in [0, 1]$:*

- *Symmetricity*

$$M(a, b) = M(b, a) \tag{6.27}$$

- *Idempotency*

$$M(a, a) = a \tag{6.28}$$

- *Monotonicity*

$$M(a, b) \leq M(c, d) \tag{6.29}$$

$$\text{if } a \leq c \text{ and } b \leq d$$

- *Boundary conditions*

$$M(0,0) = 0 \quad \text{and} \quad M(1,1) = 1 \tag{6.30}$$

- *Continuity*

$$M \text{ is continuous}$$

The following theorem proves that, whichever is the particular definition of an averaging operation, the result always falls between the standard fuzzy intersection and the standard fuzzy union.

Theorem 6.5. *If M is an averaging operator, then:*

$$\min(a,b) \leq M(a,b) \leq \max(a,b) \tag{6.31}$$

for all $a, b \in [0,1]$.

Proof:

From idempotency and monotonicity of M it follows that:

$$\min(a,b) = M(\min(a,b), \min(a,b)) \leq M(a,b)$$

and

$$M(a,b) \leq M(\max(a,b), \max(a,b)) = \max(a,b)$$

which ends the proof.

The implications of Theorem 6.5 are graphically represented in Fig. 6.3.

Similar to the other aggregation operations, different averaging operations can be defined. One class of averaging operations that covers the entire range between the standard fuzzy intersection and the standard fuzzy union is that of the *quasi-arithmetic means* M_m, defined as,

$$M_m(a,b) = f^{-1}\left(\frac{f(a) + f(b)}{2}\right) \tag{6.32}$$

The most often used mean operators are:

- Harmonic mean: $M(a,b) = 2\,a\,b/(a+b)$
- Geometric mean: $M(a,b) = \sqrt{a\,b}$
- Arithmetic mean: $M(a,b) = (a+b)/2$
- Generalized p-mean: $M(a,b) = ((a^p + b^p)/2)^{1/p}, \quad p \geq 1$

Another class of averaging operations that covers the entire range between the standard fuzzy intersection and the standard fuzzy union is called the class of ordered weighted averaging operations (OWA).

Let

$$\mathbf{w} = \langle w_1, w_2, \ldots, w_n \rangle$$

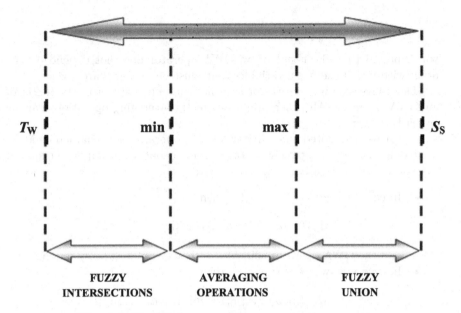

T_W min   max S_S

FUZZY   **AVERAGING**   **FUZZY**
INTERSECTIONS   **OPERATIONS**   **UNION**

Fig. 6.3. The whole range of fuzzy aggregation operation.

be a *weighting vector*, such as $w_i \in [0,1]$ for all $i \in N_n$ and

$$w_1 + w_2 + \cdots + w_n = 1$$

Then, the OWA operator M_w associated with **w** is the function

$$M_w(a_1, a_2, \ldots, a_n) = w_1\, b_1 + w_2\, b_2 + \cdots + w_n\, b_n \qquad (6.33)$$

where vector $\langle b_1, b_2, \ldots, b_n \rangle$ is a permutation of vector $\langle a_1, a_2, \ldots, a_n \rangle$ in which its elements are ordered from the largest to the smallest.

In the simplified case of $n = 2$, Eq. (6.33) becomes:

$$M_w(a,b) = w_1\, \max(a,b) + w_2\, \min(a,b) \qquad (6.34)$$

where $w_1 + w_2 = 1$.

A fundamental aspect of this operator is the reordering step. In fact, a membership grade a_i is not associated with a particular weight w_i, but a weight is associated with a particular ordered position of the membership grades $a_1, a_2, \ldots, a_n$.

To prove this statement, let us consider the following example. Let **w** = $\langle 0.4, 0.6 \rangle$ be the weighting vector, and let us consider the OWA operators $M_w(0.3, 0.5)$ and $M_w(0.5, 0.3)$, where the order of the fuzzy variables that must be averaged has been interchanged. It is

$$M_w(0.3, 0.5) = 0.4 \cdot 0.5 + 0.6 \cdot 0.3 = 0.38$$

and

$$M_w(0.5, 0.3) = 0.4 \cdot 0.5 + 0.6 \cdot 0.3 = 0.38$$

which proves that the result of an OWA operator does not depend on the order with which the fuzzy variables that must be averaged are given.

The above example clearly proves the commutativity property (6.27) of an OWA operator. All other properties of the averaging operations can be similarly proved.

It can be also noted that different OWA operators are characterized by their different weighting vectors. Yager pointed out three important special cases of OWA aggregations.

- In case $\mathbf{w} = \mathbf{w}^* = \langle 1, 0, \ldots, 0 \rangle$, then

$$M_{w^*}(a_1, a_2, \ldots, a_n) = \max\{a_1, a_2, \ldots, a_n\}$$

  and the OWA operator coincides with the standard fuzzy union.
- In case $\mathbf{w} = \mathbf{w}_* = \langle 0, 0, \ldots, 1 \rangle$, then

$$M_{w_*}(a_1, a_2, \ldots, a_n) = \min\{a_1, a_2, \ldots, a_n\}$$

  and the OWA operator coincides with the standard fuzzy intersection.
- In case $\mathbf{w} = \mathbf{w}_A = \langle 1/n, 1/n, \ldots, 1/n \rangle$, then

$$M_{w_A}(a_1, a_2, \ldots, a_n) = a_1 + a_2 + \cdots + a_n)/n$$

  and the OWA operator coincides with the arithmetic mean defined above.

As weighting vectors $\mathbf{w}^*$ and $\mathbf{w}_*$ are the upper and lower bound, respectively, of any weighting vector, hence OWA operators M_{W^*} and M_{W_*} are the boundaries of any OWA operators. That is,

$$M_{w_*} \leq M_w \leq M_{w^*}$$

To classify OWA operators with respect to their location between the standard fuzzy intersection "min" and the standard fuzzy union "max", Yager introduced a measure of "orness", where this term indicates how much the OWA operator is near the "or" operator, which is another way to specify the "max" operator. The orness is defined as follows:

$$\text{orness}(M_w) = \frac{1}{n-1} \sum_{i=1}^{n} (n-i) w_i$$

It is easy to prove that, whichever is the weighting vector $\mathbf{w}$, orness(M_w) is always in the unit interval. Moreover,

- orness$(M_{w^*}) = 1$

- orness$(M_{w_*}) = 0$
- orness$(M_{w_A}) = 0.5$

Similarly, a measure of "andness" can be defined to indicate how much the OWA operator is near the "and" operator, which is another way to specify the "min" operator. The andness is defined as follows:

$$\text{andness}(M_w) = 1 - \text{orness}(M_w)$$

Another class of averaging operators is that of the *associative averaging operations*. As their name suggests, an associative averaging operator M_a satisfies the associativity property:

$$M_a(a, M_a(b, d)) = M_a(M_a(a, b), d)$$

Indeed the associative averaging operations are a subclass of *norm operations*, which are a special kind of the aggregation operations. Norm operations h satisfy the properties of monotonicity, commutativity, and associativity of t-norms and t-conorms, but they replace the boundary conditions of t-norms and t-conorms with the following weaker boundary conditions:

$$h(0,0) = 0 \quad \text{and} \quad h(1,1) = 1$$

Due to their associativity, norm operations can be extended to any finite number of arguments.

It can be easily proven that, when a norm operation also satisfies to $h(a, 1) = a$, it becomes a t-norm; when it satisfies to $h(a, 0) = a$, it becomes a t-conorm. Otherwise, it is an associative averaging operation. An example is given by:

$$M_{a_\lambda}(a, b) = \begin{cases} \min(\lambda, S(a, b)) & \text{when } a, b \in [0, \lambda] \\ \max(\lambda, T(a, b)) & \text{when } a, b \in [\lambda, 1] \\ \lambda & \text{otherwise} \end{cases}$$

for all $a, b \in [0, 1]$, where $\lambda \in [0, 1]$, T is a t-norm, and S is a t-conorm.

6.1.4 Discussion

In the previous sections, all aggregation operators have been defined in terms of membership functions, as they are generally defined in the literature.

Let us remember that, when fuzzy variables have been defined in Chapter 2 (as well as RFVs in Chapter 4), two alternative definitions have been given. The first one is based on the membership function of the fuzzy variable, and the second one is based on its α-cuts.

It has been also stated that these two definitions are perfectly equivalent, because the α-cuts of a fuzzy variable can be readily obtained from its membership function, and its membership function can be readily obtained from its α-cuts.

However, it has been also stated that, when the fuzzy variables (and RFVs) are used to represent measurement results, the definition in terms of α-cuts is surely the better. In fact, the α-cuts of a fuzzy variable (or RFV) directly provide the confidence intervals and associated levels of confidence of the considered measurement result.

As shown in the previous sections, when two fuzzy-variables A and B, with $\mu_B(x) \leq \mu_A(x)$ for all $x \in X$, like the ones in Fig. 6.2, are considered, the aggregation operations define a new fuzzy variable. As every fuzzy variable can be defined in terms of both its membership function and its α-cuts, in this particular situation, the aggregation operations can be also alternatively defined as operating on the α-cuts of the initial fuzzy variables, instead of on their membership grades.

In other words, when two fuzzy variables, of which one is included in the other one, are considered, the aggregation operators can be alternatively defined vertically or horizontally. Vertically refers to the fact that the membership grade of the final fuzzy variable in each value $x \in X$ is determined from the membership grades of the initial fuzzy variables in the same value, and the value varies horizontally in all of the universal set X. On the other hand, horizontally refers to the fact that each α-cut at level α of the final fuzzy variable is determined from the α-cuts of the initial fuzzy variables at the same value α, and value α varies vertically from 0 to 1. This is shown in Fig. 6.4 in the case of the averaging operations: When the 'vertical' definition is considered, all values $x \in X$ must be taken into account; when the 'horizontal' definition is considered, all values $\alpha \in [0, 1]$ must be taken into account. This also suggests a more practical application of the 'horizontal' definition, with respect to the 'vertical' one.

In fact, the definition based on the α-cuts always allows the same number of computations, whichever is the universal set. On the other side, the definition based on the membership functions requires, in order to allow the same resolution, a number of computations which depends on the width of the universal set: the wider is the universal set, the greater is the number of computations required.

Figure 6.4 also shows that, when the horizontal approach is considered, the considered aggregation operation, whichever it is, must be applied twice for each considered level α.

The transformation of the definitions that have been given following a vertical approach into equivalent definitions based on an horizontal approach is not always immediate. Generally, the difficulty in finding the new definitions depends on the considered operator.

To give an example, let us consider the standard fuzzy intersection and the standard fuzzy union. Let us consider two fuzzy variables A and B, one of which is included in the other one, and let $[a_1^\alpha, a_2^\alpha]$ and $[b_1^\alpha, b_2^\alpha]$ be their generic

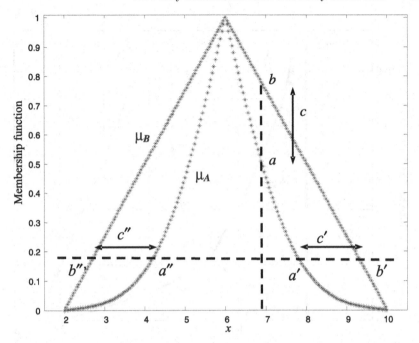

Fig. 6.4. Vertical and horizontal definitions of the averaging operations $c = M(a, b)$.

α-cuts. Then, the generic α-cuts of the standard fuzzy intersection and union are respectively given by

$$[\max\{a_1^\alpha, b_1^\alpha\}, \min\{a_2^\alpha, b_2^\alpha\}]$$

$$[\min\{a_1^\alpha, b_1^\alpha\}, \max\{a_2^\alpha, b_2^\alpha\}]$$

The proof is given by Figs. 6.1 and 6.2.

6.2 Fuzzy intersection area and fuzzy union area

The fuzzy intersection area and the fuzzy union area are based, respectively, on the definition of fuzzy intersection and fuzzy union. Hence, it could be stated that different definitions exist, based on the different kinds of t-norms and t-conorms defined in the previous section.

However, the fuzzy intersection and union area are generally defined by considering the standard fuzzy intersection and the standard fuzzy union. This is also the definition that will be used in the following chapters. Under this assumption, given two fuzzy variables A and B, it is

$$Int(A, B) = \int_{-\infty}^{+\infty} \min(A(x), B(x)) dx \qquad (6.35)$$

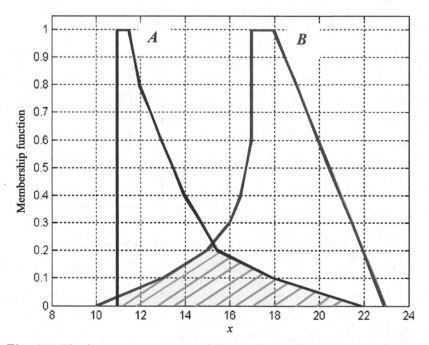

Fig. 6.5. The fuzzy intersection area between the two fuzzy variables A and B is numerically equal to the dashed area.

$$Un(A, B) = \int_{-\infty}^{+\infty} \max(A(x), B(x))dx \qquad (6.36)$$

where function "min" and "max" are the standard fuzzy intersection and standard fuzzy union, respectively, as defined in the previous section.

Examples of the fuzzy intersection and union area are reported in Figs. 6.5 and 6.6, respectively. It can be noted that the intersection area between the two fuzzy variables, which is reported in Fig. 6.5 as the dashed area, is numerically equal to the area subtended by the standard fuzzy intersection. Similarly, the union area between the two fuzzy variables, which is reported in Fig. 6.6 as the dashed area, is numerically equal to the area subtended by the standard fuzzy union.

6.3 Hamming distance

Let us consider two fuzzy variables A and B, with membership function $\mu_A(x)$ e $\mu_B(x)$, respectively. The Hamming distance between A and B is defined by

$$d(A, B) = \int_{-\infty}^{+\infty} |\mu_A(x) - \mu_B(x)|dx \qquad (6.37)$$

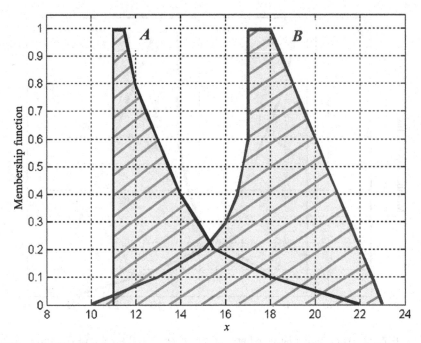

Fig. 6.6. The fuzzy union area between the two fuzzy variables A and B is numerically equal to the dashed area.

An example is reported in Fig. 6.7. It can be noted that the Hamming distance is numerically equal to the dashed area. Moreover, by comparing Fig. 6.7 with Figs. 6.5 and 6.6, it can be noted that the Hamming distance can be readily obtained by the fuzzy intersection area and union area. Hence, an equivalent definition is

$$d(A, B) = Un(A, B) - Int(A, B)$$

6.4 Greatest upper set and greatest lower set

These two operators are defined on a single fuzzy variable. Let us then consider a fuzzy variable A with membership function $\mu_A(x)$. The greatest upper set of a fuzzy variable A is still a fuzzy variable, called A^+. The membership function of A^+ is defined as

$$\mu_{A^+}(x) = \max_{y \leq x} \mu_A(y) \quad \forall x, y \in X \tag{6.38}$$

Figure 6.8 shows that the greatest upper set of a fuzzy variable is always a fuzzy variable with a nonfinite support.

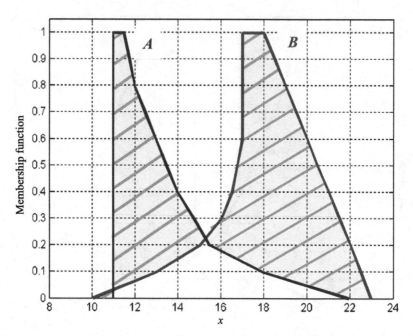

Fig. 6.7. The Hamming distance between two fuzzy variables A and B is numerically equal to the dashed area.

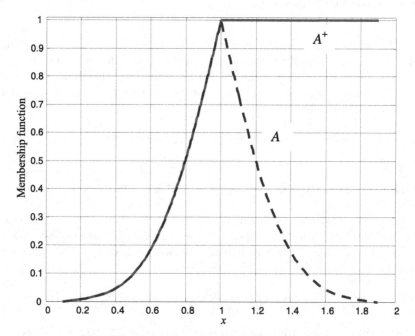

Fig. 6.8. Greatest upper set of a fuzzy variable.

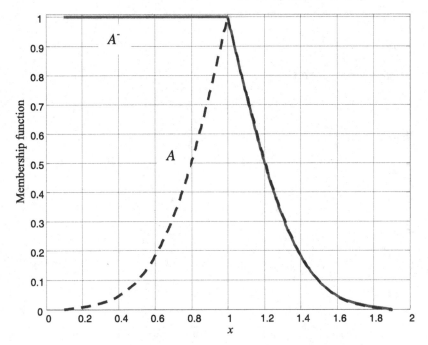

Fig. 6.9. Greatest lower set of a fuzzy variable.

As the greatest upper set of a fuzzy variable is a fuzzy variable, it is possible to define it also in terms of α-cuts. In fact, if $[a_1^\alpha, a_2^\alpha]$ is the generic α-cut of the initial fuzzy variable A, the generic α-cut of the greatest upper set A^+ is

$$[a_1^\alpha, +\infty]$$

On the other hand, the greatest lower set of the same fuzzy variable A is still a fuzzy variable, now called A^-. The membership function of A^- is defined as

$$\mu_{A^-}(x) = \max_{y \geq x} \mu_A(y) \quad \forall x, y \in X \tag{6.39}$$

Figure 6.9 shows that the greatest lower set of a fuzzy variable is always a fuzzy variable with a nonfinite support. Moreover, the generic α-cut of the greatest lower set A^- is

$$[-\infty, a_2^\alpha]$$

6.5 Fuzzy-max and fuzzy-min

Let us consider two fuzzy variables A and B, with membership function $\mu_A(x)$ and $\mu_B(x)$, respectively. The fuzzy-max operator between A and B is called

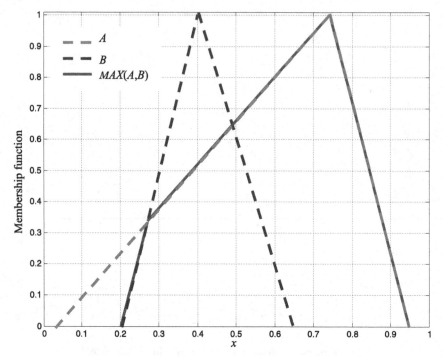

Fig. 6.10. Fuzzy-max operator between two fuzzy variables.

$\text{MAX}(A, B)$. It is still a fuzzy variable, whose membership function is defined by

$$\mu_{\text{MAX}(A,B)}(z) = \sup_{z=\max\{x,y\}} \min\{\mu_A(x), \mu_B(x)\} \quad \forall x, y, z \in X \qquad (6.40)$$

An example of fuzzy variable obtained applying the fuzzy-max operator between two fuzzy variables A and B is given in Fig. 6.10. The fuzzy variables in dashed lines are the initial variables, whereas the one in solid line is the final one, obtained by applying Eq. (6.40).

It is also possible to define the fuzzy-max operator in terms of α-cuts. In fact, if $[a_1^\alpha, a_2^\alpha]$ and $[b_1^\alpha, b_2^\alpha]$ are the generic α-cuts of the initial fuzzy variables A and B, the generic α-cut of the fuzzy-max is

$$[\max\{a_1^\alpha, b_1^\alpha\}, \max\{a_2^\alpha, b_2^\alpha\}]$$

for every $\alpha \in [0, 1]$.

The fuzzy-min operator between A and B is called $\text{MIN}(A, B)$. It is still a fuzzy variable, whose membership function is defined by:

$$\mu_{\text{MIN}(A,B)}(z) = \inf_{z=\max\{x,y\}} \min\{\mu_A(x), \mu_B(x)\} \quad \forall x, y, z \in X \qquad (6.41)$$

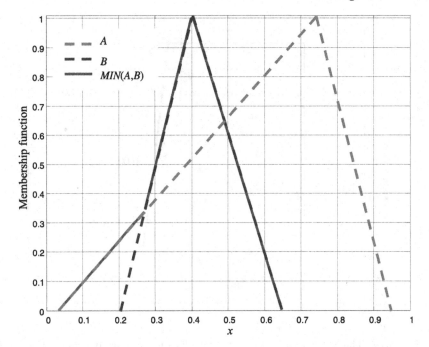

Fig. 6.11. Fuzzy-min operator between two fuzzy variables.

An example of fuzzy variable obtained applying the fuzzy-min operator between two fuzzy variables A and B is given in Fig. 6.11. The fuzzy variables in dashed lines are the initial variables, whereas the one in solid line is the final one, obtained by applying Eq. (6.41).

It is also possible to define the fuzzy-min operator in terms of α-cuts. The generic α-cut of the fuzzy-min variable is given by

$$[\min\{a_1^\alpha, b_1^\alpha\}, \min\{a_2^\alpha, b_2^\alpha\}]$$

for every $\alpha \in [0, 1]$.

6.6 Yager area

The Yager area represents, in a certain sense, the distance of a fuzzy variable from the origin. Given a fuzzy variable A, with membership function $\mu_A(x)$, the Yager area is determined as

$$Y_A = \int_0^1 y_A(\alpha)d\alpha \tag{6.42}$$

where $y_A(\alpha)$ is the function of the middle points of the various α-cuts at the different levels α.

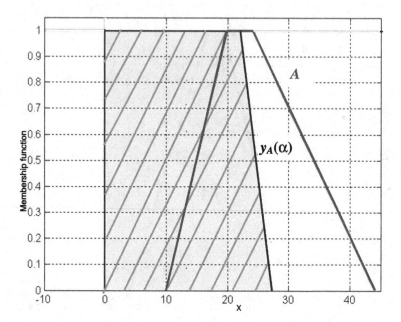

Fig. 6.12. The Yager area of the fuzzy variable A is numerically equal to the dashed area.

An example of evaluation of the Yager area is reported in Fig. 6.12. The Yager area is numerically equal to the dashed area in the figure. It can be noted that the more the fuzzy variable is distant from the origin, the greater is the value assumed by this quantity.

7

The Mathematics of Random–Fuzzy Variables

In the previous chapters, the random–fuzzy variables have been defined and proved to be suitable for the representation of measurement results, together with their associated uncertainty.

It has been also underlined that RFVs can be successfully employed, provided that a suitable probability–possibility transformation and a suitable mathematics are defined. Chapter 5 showed how to meet the first requirement, whereas the second one will be discussed in this chapter.

As an RFV is a specific fuzzy variable of type 2 and as a proper mathematics has been already defined for fuzzy variables of type 1, the first idea could be simply to extend this last mathematics. The extension would be immediate. In fact, as a fuzzy variable of type 1 is defined by a membership function, and a fuzzy variable of type 2 is defined by two membership functions, the mathematics of fuzzy variables could be simply applied twice: a first time for the internal membership functions and the second time for the external membership functions of the RFVs to be processed.

Unfortunately, this simple approach does not fulfill the main requirement for an RFV: Its α-cuts are expected to provide the confidence intervals within which the value that could be reasonably attributed to the measurand is supposed to lie, for all possible levels of confidence between 0 and 1. In particular, the two subintervals in each α-cut, located between the internal and external membership functions, are expected to represent the dispersion of the measurement results due to random phenomena.

However, as widely discussed in Chapter 2, the mathematics of fuzzy variables can represent only the behavior of systematic contributions, when they propagate through the measurement process. Hence, the application of this mathematics on both the internal and the external membership functions of an RFV would prevent one from correctly accounting for the behavior of random contributions, when they propagate through the measurement process.

The above considerations lead to conclude that a different kind of mathematics is needed. Some useful, intuitive hints for the definition of such a mathematics can be retrieved from a brief analysis of the way random contributions combine together, according to the fundamentals of the probability theory.

7.1 Combination of the random contributions

It is well known that, when different contributions to uncertainty combine together, the final result is strongly dependent on their behavior: Random contributions lead to different results from those originated by systematic contributions or unknown contributions. This is perfectly consistent with the assumption that the combination of two measurement results will provide a different amount of information, depending on the available information associated with the uncertainty contributions affecting the results.

Let us suppose, for instance, that, because of the uncertainty contributions, two measurement results fall in intervals [1.5, 2.5] and [3, 4], respectively, and let us suppose that the final measurement result is obtained as their sum. Then, different situations can be met.

If the available information suggests that systematic contributions are affecting measurement results, or there is no further available information to assess the nature of the contributions, then according to the Theory of Evidence, the two measurement results must be represented by rectangular possibility distributions. In this case, when the sum of the two results is taken into account, the mathematics of the fuzzy variables must be applied, and the result is still represented by a rectangular possibility distribution (see Chapter 2). This is shown in Fig. 7.1, from which it can be also noted that the width of the support of the possibility distribution of the result is the sum of the widths of the supports of the initial possibility distributions.

On the other hand, if the available information suggests that the two measurement results are affected by uniformly distributed random contributions to uncertainty, then if the probability theory is applied, they can be represented by rectangular probability distributions, and their height is of course evaluated by considering that the probability of all events (their subtended area) must be always 1.

In this case, different results can be obtained, depending on the degree of correlation of the two contributions. If, for example, the two contributions are totally uncorrelated, the probability distribution function of the result of the sum is the convolution of the two initial distributions. In the considered example, this leads to a triangular probability distribution, as shown in Fig. 7.2. Also in this case, it can be stated that the width of the support of the probability distribution of the result is given by the sum of the widths of the supports of the initial probability distributions.

However, this last statement applies only to the particular case of totally uncorrelated random contributions. For a more general analysis, let us consider the case of correlated contributions.

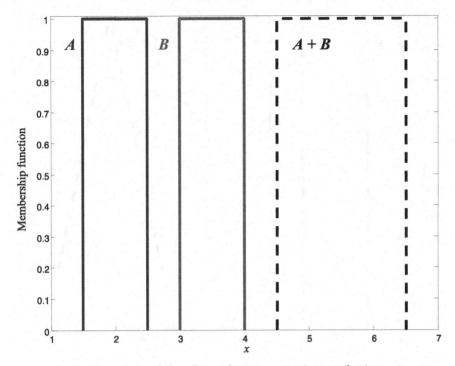

Fig. 7.1. Sum of the effects of two systematic contributions.

From a theoretical point of view, it is well known that the correlation between two probability distributions X and Y can be quantified by the correlation coefficient:

$$\rho(X,Y) = \frac{u(X,Y)}{u(X)\,u(Y)}$$

where $u(X,Y)$ is the covariance of the two distributions and $u(X)$ and $u(Y)$ are the standard deviations of X and Y, respectively.

Therefore, $\rho(X,Y)$ ranges from -1 to $+1$, where $\rho(X,Y) = -1$ means total negative correlation, $\rho(X,Y) = 0$ means total uncorrelation, and $\rho(X,Y) = 1$ means total positive correlation.

These three values are the most commonly used values in the measurement practice, because these are the three most frequently met situations. For this reason, and because they are also the three limit situations, they will be considered in the following discussion.

The effect of a zero correlation coefficient has already been considered in the example of Fig. 7.2: All values in the two considered intervals can combine randomly, thus leading to the triangular probability distribution, which shows how the central value is more probable because it can be obtained by different couples of values.

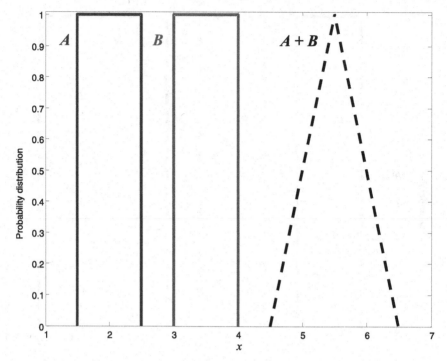

Fig. 7.2. Sum of the effects of two uncorrelated random contributions.

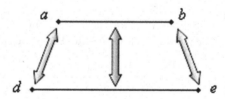

Fig. 7.3. The effect of a positive total correlation.

On the other hand, total correlation (both positive and negative) acts in the following way: When a random extraction from the first interval is taken, then the extraction in the second interval is fully determined. This means that the values in the two intervals cannot randomly combine, and only some couples of values can be extracted, as schematically shown in Figs. 7.3 and 7.4. When the two initial uniformly distributed measurement results are considered, this leads to the probability distributions shown in Figs. 7.5 and 7.6, for correlation coefficients +1 and −1, respectively.

Figure 7.5 shows that, when total positive correlation is considered, and the initial distributions have the same shape, the shape of the probability distribution of the result is the same as that of the initial probability distributions. In the particular case of Fig. 7.5, the result distributes according to

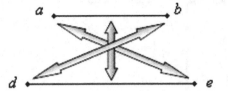

Fig. 7.4. The effect of a positive negative correlation.

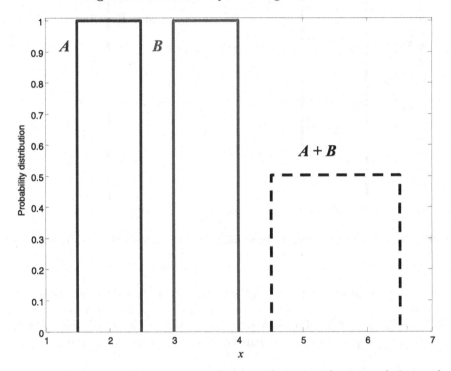

Fig. 7.5. Sum of the effects of two random contributions with +1 correlation coefficient.

a rectangular probability distribution. In fact, a certain value of the result can be obtained only by summing up the values in a specific possible pair, as shown in Fig. 7.3. Moreover, the support of the result is the sum of the supports of the two initial distributions. In fact, due to the positive total correlation, the extraction of one edge (left or right) of the first considered interval is surely coupled with the same edge of the second interval.

On the other hand, Fig. 7.6 shows that, when total negative correlation is considered, and the initial distributions have the same shape and width, the result obtained in the considered example is a single scalar value. In fact, due to the negative total correlation, the extraction of one edge (left or right) of the first considered interval is surely coupled with the opposite edge of the

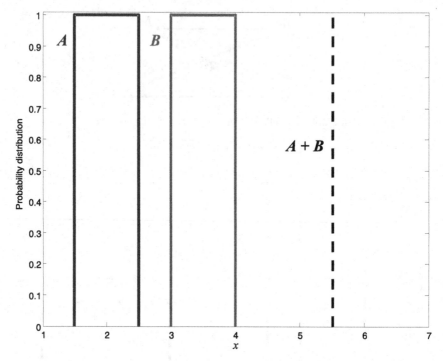

Fig. 7.6. Sum of the effects of two random contributions with −1 correlation coefficient.

second interval, the extraction of the middle point of the first interval is surely coupled with the same point in the second interval, and so on, as schematically shown in Fig. 7.4. If all possible pairs are considered, it can be concluded that the sum of the values in each pair is always the same (5.5 in the example of Fig. 7.6).

Let us consider, however, that the single scalar value is a degeneration of a rectangular probability distribution, for which the width of the support is zero, and the height is, by definition, infinite. This particular result is obtained because the two initial distributions have the same width of the supports. In the more general case, where the two effects to be combined are represented by probability distributions with different widths of the supports, the width of the support of the final result is the difference of the two initial widths.

When a correlation coefficient different from −1, 0, or +1 is taken, the way two random variables combine cannot be a priori determined in the same easy way as that of the above examples, because both the support and the shape of the final distribution depend on the given correlation coefficient.

The above example about the sum of two random variables showed that, as far as the width of the final support is concerned, it can vary from a minimum to a maximum value, where the minimum value is the difference of the widths of the two initial supports, and the maximum value is the sum of the same widths. It can be expected that different values of the correlation factors lead to intermediate results.

The same example also showed that, as far as the shape of the final distribution is concerned, it can be the same as the initial ones, or different. In the considered example, where the initial distributions were rectangular, the final distributions were again rectangular, in case of total correlation, or triangular, in case of uncorrelation. Different shapes can be expected for different values of the correlation factors.

The following sections will try to generalize the above conclusions, as far as possible.

7.1.1 Supports

Let us now consider two generic probability distributions, no matter what are their shapes, and let $[a, b]$ and $[d, e]$ be their supports.

The aim of this section is to determine, for each of the four arithmetic operations, the support of the final distribution.

This can be done, in a strict mathematical way, only for the particular cases with $\rho = 0$, $\rho = +1$, and $\rho = -1$, which are, as stated above, the most interesting cases in the measurement field. In all other cases, when the support cannot be strictly determined in closed form, either a Monte Carlo simulation can be performed or interpolated values can be obtained from those determined strictly.[1]

Sum

Let us consider the sum of the two given distributions. When $\rho = 0$ is considered, all values in interval $[a, b]$ can combine randomly with all values in interval $[d, e]$. Therefore, the support of the final distribution is given by

$$[a + d, b + e] \tag{7.1}$$

which considers all possible couples of values from the above considered intervals.

When $\rho = +1$ is considered, the values in the two intervals can only combine according to Fig. 7.3. This leads again to a final support:

$$[a + d, b + e]$$

This also leads to the conclusion that a positive correlation does not change the support defined in the case of total uncorrelation. Therefore, Eq. (7.1) can be taken as the support of the final distribution for every ρ, for $0 \leq \rho \leq 1$.

---

[1] Finding an approximation, instead of a strict solution, is somehow acceptable if we consider that, in most practical applications, ρ values different from 0, +1, or −1 are often a first approximation estimate and are seldom evaluated from the analysis of the covariance matrix.

When $\rho = -1$ is considered, the values in the two intervals can only combine according to Fig. 7.4. This leads to the final support:

$$[\min\{a + e; b + d\}, \max\{a + e; b + d\}] \tag{7.2}$$

Let us note that, in the case in which the two considered intervals have the same width, it is

$$b - a = e - d \tag{7.3}$$

As the above relationship can be rewritten as

$$b + d = a + e$$

it follows that interval (7.2) degenerates into a single value, as also shown in Fig. 7.6.

Difference

Let us consider the difference of the two given distributions.

When $\rho = 0$ is considered, all values in interval $[a, b]$ can combine randomly with all values in interval $[d, e]$. Therefore, the support of the final distribution is given by

$$[a - e, b - d] \tag{7.4}$$

which considers all possible couples of values from the above considered intervals.

When $\rho = +1$ is considered, the values in the two intervals can only combine according to Fig. 7.3. This leads to a final support:

$$[\min\{a - d; b - e\}, \max\{a - d; b - e\}] \tag{7.5}$$

Let us note that, in the case in which the two considered intervals have the same width, relationship (7.3) applies, and as it can be also rewritten as

$$b - e = a - d$$

it follows that interval (7.5) degenerates into a single value.

When $\rho = -1$ is considered, the values in the two intervals can only combine according to Fig. 7.4. This leads again to the final support in Eq. (7.4). It can be concluded that a negative correlation does not change the support defined in the case of total uncorrelation. Therefore, Eq. (7.4) can be taken as the support of the final distribution for every ρ, for $-1 \leq \rho \leq 0$.

Product

Let us consider the product of the two given distributions.

When $\rho = 0$ is considered, all values in interval $[a, b]$ can combine randomly with all values in interval $[d, e]$. Therefore, the support of the final distribution is given by

$$[\min\{ad, ae, bd, be\}, \max\{ad, ae, bd, be\}] \tag{7.6}$$

which considers all possible couples of values from the above considered intervals.

When $\rho = +1$ is considered, the values in the two intervals can only combine according to Fig. 7.3. Therefore, the final support $[c_1, c_2]$ is given by

$$c_1 = \begin{cases} x_m \, y_m & \text{if } a < x_m < b \\ \min\{ad, be\} & \text{otherwise} \end{cases} \tag{7.7}$$

$$c_2 = \max\{ad, be\}$$

where, provided that $a < x_m < b$, x_m and y_m are, respectively, the extractions in intervals $[a, b]$ and $[d, e]$, which lead to a minimum in the product function. In particular, it is

$$x_m = \frac{\mu_1 - \mu_2 \, r}{2} \tag{7.8}$$

and

$$y_m = \mu_2 + \frac{x_m - \mu_1}{r} \tag{7.9}$$

where μ_1 and μ_2 are, respectively, the mean values of intervals $[a, b]$ and $[d, e]$, and r is the ratio between the widths of the two intervals. It is

$$\mu_1 = \frac{a + b}{2}$$

$$\mu_2 = \frac{d + e}{2}$$

$$r = \frac{b - a}{e - d}$$

Proof of (7.7):

Let us consider intervals $[a, b]$ and $[d, e]$ and a total positive correlation between the two distributions. Then, a generic extraction x from interval $[a, b]$ determines an extraction y_{+1} from the second interval; according to Fig. 7.3, it is

$$y_{+1} = \mu_2 + \frac{x - \mu_1}{r} \tag{7.10}$$

Let $p_{+1}(x)$ be the product function between all values in the possible couples. Hence,

$$p_{+1}(x) = x \, y_{+1} = \frac{x^2}{r} + x \left(\mu_2 - \frac{\mu_1}{r} \right)$$

It can be recognized that $p_{+1}(x)$ is the equation of a parabola. Moreover, being $r > 0$ by definition, this parabola is up-concave and hence presents an absolute minimum in correspondence with its vertex.

The first derivative of $p_{+1}(x)$ is

$$p'_{+1}(x) = 2\frac{x}{r} + \left(\mu_2 - \frac{\mu_1}{r}\right)$$

which is zero in

$$x_m = \frac{\mu_1 - \mu_2\,r}{2}$$

If $a < x_m < b$, this proves Eq. (7.8).

Equation (7.9) is also proved, by replacing Eq. (7.8) in Eq. (7.10).

If $x_m \notin (a,b)$ and, of course, $y_m \notin (d,e)$, it can be readily checked that the minimum of $p_{+1}(x)$ is given by either product ad or product be. This proves Eq. (7.7).

In particular, it can be also stated that the minimum of $p_{+1}(x)$ is given by product ad if $x_m < a$, and by product be if $b < x_m$. In fact, in the two cases, the left and right arcs of the parabola are considered, respectively.

When $\rho = -1$ is considered, the values in the two intervals can only combine according to Fig. 7.4. This leads again to the final support $[c_1, c_2]$, where

$$c_1 = \min\{ae, bd\}$$

$$c_2 = \begin{cases} x_M\,y_M & \text{if } a < x_M < b \\ \max\{ae, bd\} & \text{otherwise} \end{cases} \tag{7.11}$$

where, provided that $a < x_M < b$, x_M and y_M are, respectively, the extractions in intervals $[a, b]$ and $[d, e]$ which lead to a maximum in the product function. In particular, it is

$$x_M = \frac{\mu_1 + \mu_2\,r}{2} \tag{7.12}$$

and

$$y_M = \mu_2 - \frac{x_M - \mu_1}{r} \tag{7.13}$$

where μ_1 and μ_2 are again, respectively, the mean values of intervals $[a, b]$ and $[d, e]$, and r is the ratio between the widths of the two intervals.

Proof of Eq. (7.11):

Let us consider intervals $[a, b]$ and $[d, e]$ and a total negative correlation between the two distributions. Then, a generic extraction x from interval $[a, b]$ determines an extraction y_{-1} from the second interval; according to Fig. 7.4, it is

$$y_{-1} = \mu_2 - \frac{x - \mu_1}{r} \tag{7.14}$$

Let $p_{-1}(x)$ be the product function between all values in the possible couples. Hence,

$$p_{-1}(x) = x\,y_{-1} = -\frac{x^2}{r} + x\left(\mu_2 + \frac{\mu_1}{r}\right)$$

It can be recognized that $p_{-1}(x)$ is the equation of a parabola. Moreover, being $r > 0$ by definition, this parabola is down-concave and hence presents an absolute maximum in correspondence with its vertex.
The first derivative of $p_{-1}(x)$ is

$$p'_{-1}(x) = -2\,\frac{x}{r} + \left(\mu_2 + \frac{\mu_1}{r}\right)$$

which is zero in

$$x_M = \frac{\mu_1 + \mu_2\,r}{2}$$

If $a < x_M < b$, this proves Eq. (7.12).
Equation (7.13) is also proved, by replacing Eq. (7.12) in Eq. (7.14).
If $x_M \notin (a,b)$ and, of course, $y_M \notin (d,e)$, it can be readily checked that the maximum of $p_{-1}(x)$ is given by either product ae or product bd. This proves Eq. (7.11).
In particular, it can be also stated that the maximum of $p_{-1}(x)$ is given by product ae if $x_M < a$, and by product bd if $b < x_M$. In fact, in the two cases, the left and right arcs of the parabola are considered, respectively.

Division

Let us consider the division of the two given distributions. Of course, in this case, this operation can be performed only if interval $[d, e]$ does not include the zero value; that is,

$$0 \notin [d, e]$$

When $\rho = 0$ is considered, all values in interval $[a, b]$ can combine randomly with all values in interval $[d, e]$. Therefore, the support of the final distribution is given by

$$[\min\{a/d, a/e, b/d, b/e\}, \max\{a/d, a/e, b/d, b/e\}] \qquad (7.15)$$

which considers all possible couples of values from the above considered intervals.
When $\rho = +1$ is considered, the values in the two intervals can only combine according to Fig. 7.3. This leads to a final support:

$$[\min\{a/d; b/e\}, \max\{a/d; b/e\}] \qquad (7.16)$$

Proof of Eq. (7.16):

The reader could ask why something similar to Eq. (7.7) does not apply. This can be easily proved. Let x be a generic extraction from interval $[a, b]$ and y_{+1} its correspondent value, evaluated by applying Eq. (7.10). Then, the division function $d_{+1}(x)$ between the values in all possible couples is given by

$$d_{+1}(x) = \frac{x}{y_{+1}} = \frac{x\,r}{x + (\mu_2\,r - \mu_1)}$$

It can be readily recognized that $d_{+1}(x)$ is the equation of an homographic function, which describes an equilateral hyperbola referred to its asymptotis. As this function is always decreasing (or always increasing[2]), no finite maximum or minimum values are present.

When $\rho = -1$ is considered, the values in the two intervals can only combine according to Fig. 7.4. This leads again to the final support:

$$[\min\{a/e; b/d\}, \max\{a/e; b/d\}] \tag{7.17}$$

Proof of Eq. (7.17):

Similarly to the proof of Eq. (7.16), let us consider the division function $d_{-1}(x)$:

$$d_{-1}(x) = \frac{x}{y_{-1}} = \frac{x\,r}{-x + (\mu_2\,r + \mu_1)}$$

It can be readily recognized that $d_{-1}(x)$ is again the equation of an homographic function. Hence, no finite maximum or minimum values are present.

7.1.2 Distributions

Once the supports of the resulting distributions have been determined, it is necessary to also evaluate the probability distribution functions of the results. Unfortunately, this task is not as immediate as the determination of the supports.

In fact, it is well known that the probability distribution functions of the sum of two uncorrelated random variables are given by the convolution of the probability distribution functions of the two variables. It is also known that, if the hypothesis of the Central Limit Theorem is met, the final probability distribution function of the sum of a large number of uncorrelated random variables tends to the normal (or Gaussian) distribution, as shown in Chapter 1. On the other hand, nothing can be a priori stated if the initial variables are somehow correlated.

The problem becomes even more complex when the product is considered. In fact, the only well-known theoretical result can only be applied to the product of two uncorrelated random variables with the same normal probability distribution functions, and it states that the probability distribution function of the product is a chi-square distribution with 2 degrees of freedom.

Outside the above particular cases, no analytical solutions showing general validity can be given. To define a mathematics of the RFV, able to at least approximate the way random variables combine together, it is important to analyze this problem from a numerical point of view, by means of Monte Carlo simulations.

---

[2] This depends on whether the hyperbola lies in the first and third quadrants, or in the second and fourth quadrants, where the quadrants are referred to the asymptotis instead of the xy axis.

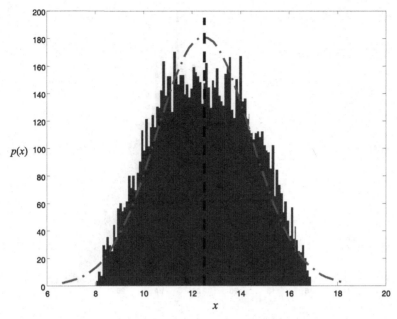

Fig. 7.7. Sum of two uncorrelated ($\rho = 0$) uniform probability distributions. The dashed line shows the sum of the means of the initial distributions. The dash-dotted line shows the normal distribution that best fits the obtained histogram.

Therefore, the following sections show some simulation results, in typical situations, for which, whenever possible, empirical rules can be extracted which approximates the behavior of random variables. Of course, if a more accurate solution is needed in some particular cases, the probability distribution function of the result can be numerically estimated by means of suitable Monte Carlo simulations and the corresponding possibility distribution can be obtained by applying the method shown in Chapter 5[3].

Sum

Let us consider two uniform uncorrelated probability distributions P_1 and P_2, over intervals [6,12] and [2,5], respectively. Figure 7.7 shows the histogram of the sum of the two distributions, obtained by applying the Monte Carlo simulation. To simulate the total uncorrelation between the initial distributions, 10,000 random extractions have been considered for each distribution, and then summed up.

---

[3] The major drawback of this approach is the heavy computational burden, much heavier than that of the RFV mathematics, as it will be clear in the following.

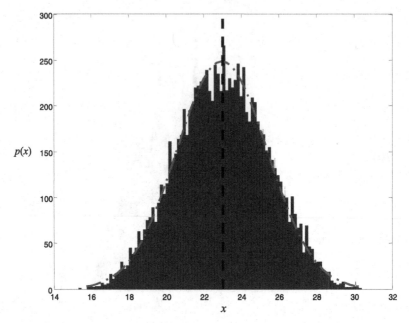

Fig. 7.8. Sum of five uncorrelated ($\rho = 0$) uniform probability distributions. The dashed line shows the sum of the means of the initial distributions. The dash-dotted line shows the normal distribution that best fits the obtained histogram.

In the same figure, the dashed line shows the sum of the means of the initial distributions, whereas the dash-dotted line shows the normal distribution that best fits the obtained histogram.

The obtained histogram is symmetric, and its mean is indeed the sum of the means of the initial distributions. The support of the histogram is interval [8,17], which confirms the theoretical expectation, given by Eq. (7.1). Moreover, the shape of the histogram approximates quite well a trapezoidal distribution, which is the expected theoretical result. This trapezoidal distribution can be also seen as a first, rough approximation of a normal distribution. This approximation refines when more than two uniform distributions are summed up, as shown in Figs. 7.8 and 7.9.

In fact, Fig. 7.8 shows the sum of five uniform distributions: The first one is the same as P_1 and distributes over interval [6,12], whereas the other four are the same as P_2 and distribute over the same interval [2,5]. A Monte Carlo simulation has been performed by considering 10,000 random extractions for each distribution.

The obtained histogram is again symmetric, its mean is the sum of the means of the initial distributions (dashed line), and its shape approximates a normal distribution (dash-dotted line) quite well.

The theoretical expected value for the support of the final distribution is [14,32]. From Fig. 7.8, the support of the obtained histogram seems to be a bit

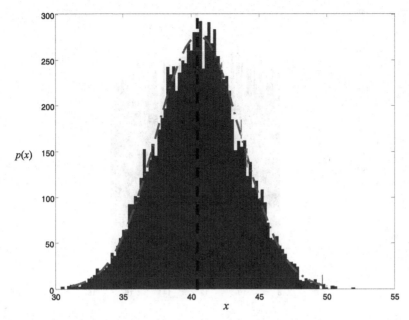

Fig. 7.9. Sum of ten uncorrelated ($\rho = 0$) uniform probability distributions. The dashed line shows the sum of the means of the initial distributions. The dash-dotted line shows the normal distribution that best fits the obtained histogram.

smaller. Let us consider, however, that the edges of the theoretical support stay on the tails of the normal distribution. This means that, also from a theoretical point of view, only a small amount of the results fall at the edges of interval [14,32].

Figure 7.9 shows the sum of ten uniform distributions: The first one is again the same as P_1 and distributes over interval [6,12], whereas the other nine are again the same as P_2 and distribute over the same interval [2,5].

The same considerations as in the previous example can be drawn. A symmetric histogram, whose mean is the sum of the means of the initial distributions (dashed line) and whose shape well approximates a normal distribution (dash-dotted line) is obtained.

The theoretical support of the final distribution is [24,57]. Even if the support of the obtained histogram seems to be a bit smaller, the same considerations drawn for the support for the histogram of Fig. 7.8 can be drawn.

The reported examples show how the Monte Carlo simulations confirm the expectations of the Central Limit Theorem: The sum of uncorrelated random variables tends to a normal distribution.

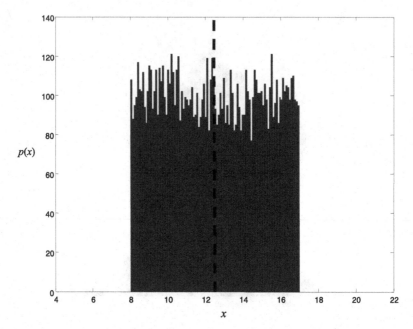

Fig. 7.10. Sum of two totally positive correlated ($\rho = +1$) uniform probability distributions. The dashed line shows the sum of the means of the initial distributions.

Let us now consider the same distributions as those taken for the examples of Figs. 7.7, 7.8, and 7.9, respectively. However, instead of considering them totally uncorrelated, let us now suppose that they are totally correlated.

As stated, total correlation can lead to two situations: a positive or a negative total correlation, which corresponds to a correlation coefficient $\rho = +1$ or $\rho = -1$, respectively.

Figures 7.10 to 7.12 show the results obtained, by applying a Monte Carlo simulation, when a correlation coefficient $\rho = +1$ is considered; Figs. 7.13 to 7.15 show the results obtained when $\rho = -1$ is considered. In particular, Figs. 7.10 and 7.13 show the results obtained when two distributions are summed up. Figures 7.11 and 7.14 show the results obtained when five distributions are summed up. Figures 7.12 and 7.15 show the results obtained when ten distributions are summed up.

All simulations have been performed as follows: 10,000 random extractions have been taken from the first interval, whereas the correspondent values in the other intervals have been determined by considering the total correlation, as shown in Figs. 7.3 and 7.4.

First of all, it can be noted that the shapes of the obtained histograms no longer approximate a normal distribution. On the other hand, they approximate a uniform distribution quite well, whichever is the number of the initial distributions that are summed up.

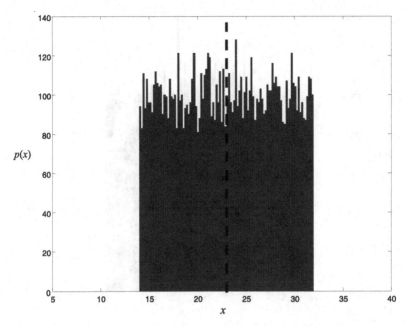

Fig. 7.11. Sum of five totally positive correlated ($\rho = +1$) uniform probability distributions. The dashed line shows the sum of the means of the initial distributions.

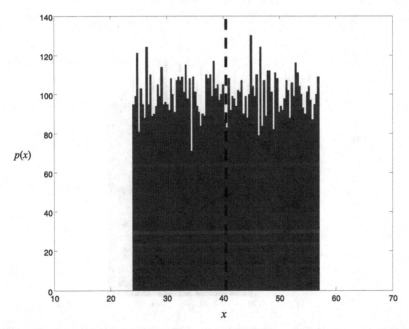

Fig. 7.12. Sum of ten totally positive correlated ($\rho = +1$) uniform probability distributions. The dashed line shows the sum of the means of the initial distributions.

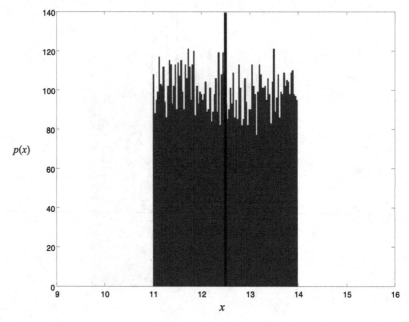

Fig. 7.13. Sum of two totally negative correlated ($\rho = -1$) uniform probability distributions. The dashed line shows the sum of the means of the initial distributions.

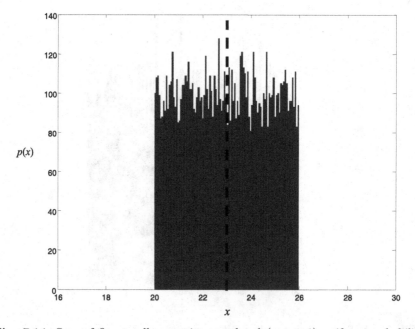

Fig. 7.14. Sum of five totally negative correlated ($\rho = -1$) uniform probability distributions. The dashed line shows the sum of the means of the initial distributions.

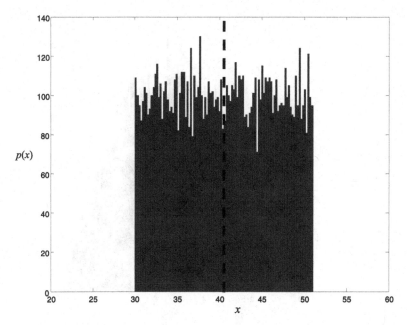

Fig. 7.15. Sum of ten totally negative correlated ($\rho = -1$) uniform probability distributions. The dashed line shows the sum of the means of the initial distributions.

This result is coherent with the fact that the initial distributions are uniform and that, when a total correlation is considered, the possible values in the considered intervals do not couple in a random way, but in a deterministic way, as shown in Figs. 7.3 and 7.4. Hence, in this considered example, the final distribution has the same shape as the initial ones.[4] Moreover, the mean values of the obtained histograms correspond to the sum of the means of the initial distributions (dashed line); in other words, the obtained histograms are symmetric with respect to this last value.[5]

As far as the supports are concerned, Eqs. (7.1) and (7.2) lead to the following results, for the given example.

Let us first consider the correlation coefficient $\rho = +1$; then, when two distributions are considered, support [8,17] is obtained; when five distributions are considered, support [14,32] is obtained; and when ten distributions are considered, support [24,57] is obtained. These theoretical supports are all confirmed by Figs. 7.10–7.12.

On the other hand, if a correlation coefficient $\rho = -1$ is applied, then, when two distributions are considered, support [11,14] is obtained; when five

---

[4] Of course, when different probability distributions are summed up, the shape of the final distribution will depend on the shapes of the two initial ones.

[5] This applies because the initial distributions are symmetric with respect to their mean values. It is obvious that, if the initial distributions were not symmetric, a dissymmetry would also be found in the final result.

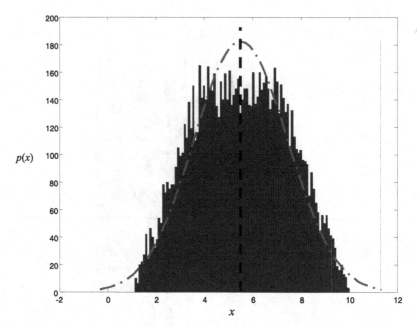

Fig. 7.16. Difference of two uncorrelated ($\rho = 0$) uniform probability distributions. The dashed line shows the difference of the means of the initial distributions. The dash-dotted line shows the normal distribution that best fits the obtained histogram.

distributions are considered, support [20,26] is obtained; and when ten distributions are considered, support [30,51] is obtained. These theoretical supports are all confirmed by Figs. 7.13–7.15.

Difference

Let us now consider the difference of the two uniform distributions P_1 and P_2 defined in the previous section.

Figure 7.16 shows the obtained histogram, when a Monte Carlo simulation (by means of 10,000 extractions) is performed, under the assumption of total uncorrelation of the two distributions.

It can be noted that the obtained histogram is symmetric with respect to the difference of the means of the initial distributions (dashed line).

The support of the obtained histogram is [1,10], which confirms the theoretical expectation, given by Eq. (7.4).

As expected, the shape of the obtained histogram is very similar to that in Fig. 7.7, where the sum of the same distributions is considered. In fact, sum and difference behave in the same way and thus lead to similar results. Therefore, the same conclusions can be drawn here, and even if no further examples are given, for the sake of brevity, it can be stated that the Central

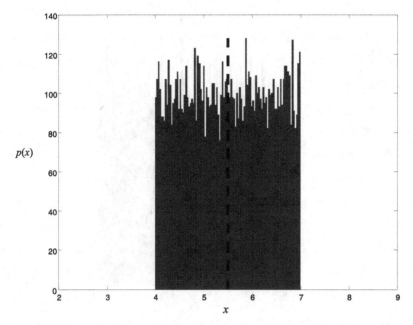

Fig. 7.17. Difference of two totally positive correlated ($\rho = +1$) uniform probability distributions. The dashed line shows the difference of the means of the initial distributions.

Limit Theorem can be applied and the difference of uncorrelated random variables always tends to a normal distribution.

Figure 7.17 shows the histogram obtained from a Monte Carlo simulation under the assumption of total positive correlation between the two considered distributions. It can be noted that the result distributes according to a uniform distribution. Even if no further examples are given, for the sake of brevity, this allows one to conclude that, in this example, the final distribution has the same shape as the initial ones,[6] as in the case of the sum operation.

The obtained histogram is again symmetric with respect to the difference of the means of the initial distributions (dashed line), and its support is [4,7], which once again confirms the theoretical expectation, given by Eq. (7.5).

Similar considerations can be also done for the example reported in Fig. 7.18, that is, the histogram obtained from a Monte Carlo simulation under the assumption of total negative correlation.

---

[6] Of course, when the difference of two different probability distributions is considered, the shape of the final distribution will depend on the shapes of the initial ones.

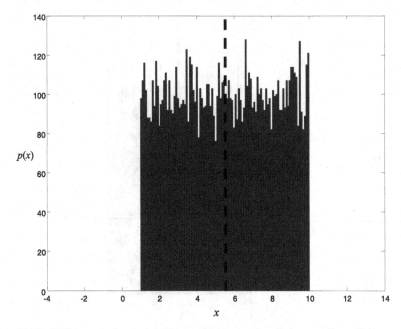

Fig. 7.18. Difference of two totally negative correlated ($\rho = -1$) uniform probability distributions. The dashed line shows the difference of the means of the initial distributions.

Product

Let us now consider the product of the two uniform distributions P_1 and P_2 defined above.

Figure 7.19 shows the obtained histogram, when a Monte Carlo simulation (by means of 10,000 extractions) is performed, under the assumption of total uncorrelation of the two distributions. Figure 7.20 shows the histogram obtained from a Monte Carlo simulation under the assumption of total positive correlation between the two considered distributions, whereas Fig. 7.21 shows the histogram obtained from a Monte Carlo simulation under the assumption of total negative correlation.

In all figures, the dashed line represents the product of the means of the initial distributions.

As a first conclusion, it can be stated that the symmetry is totally lost. In fact, the obtained distributions are no longer symmetric (even if the initial ones were), and the product of the means of the initial distributions no longer falls in the middle of the supports of the obtained histograms.

This is not the only difference from the results obtained for the sum and difference operations. In fact, as shown in Fig. 7.19, when uncorrelated random variables are considered, the shape of the obtained histogram can be no longer

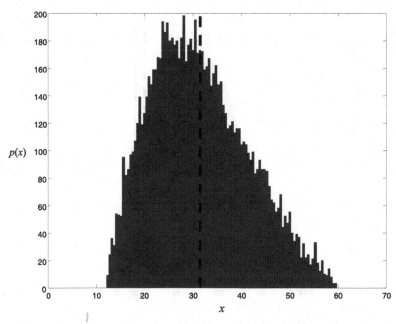

Fig. 7.19. Product of two uncorrelated ($\rho = 0$) uniform probability distributions. The dashed line shows the product of the means of the initial distributions.

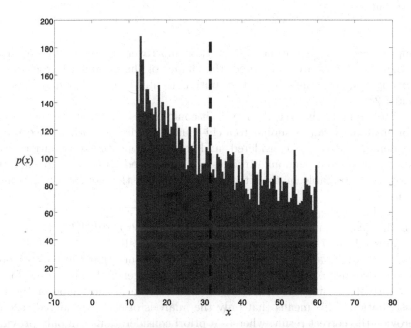

Fig. 7.20. Product of two totally positive correlated ($\rho = +1$) uniform probability distributions. The dashed line shows the product of the means of the initial distributions.

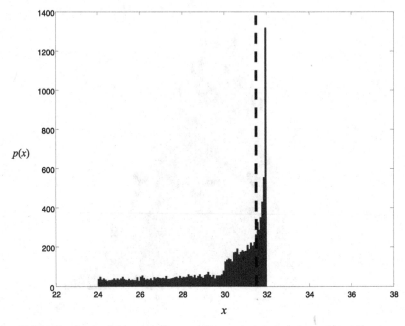

Fig. 7.21. Product of two totally negative correlated ($\rho = -1$) uniform probability distributions. The dashed line shows the product of the means of the initial distributions.

approximated by a normal distribution; moreover, when totally correlated random variables are considered, the shapes of the obtained histograms are no longer equal to the shapes of the initial distributions, as shown by Figs. 7.20 and 7.21.

From an intuitive visual analysis, it could be assessed that the shape of the obtained histogram is similar to a chi-square distribution, when uncorrelated random variables are considered, and to an exponential distribution, when totally correlated random variables are considered. However, no analytical support is available in the literature to support this conclusion in general situations.

As far as the supports are concerned, Eqs. (7.6), (7.7), and (7.11) lead, for the given example, to intervals [12,60], [12,60], and [24,32], respectively. Figures 7.19 to 7.21 confirm the theoretical intervals.

These observations all provide empirical evidence that the product operation is complex. An analytical solution can be generally given only for the support of the final distribution, but not for the actual shape of the final distribution. This means that only the analysis of each considered case can provide the correct result, whereas a priori considerations can only provide a first approximation.

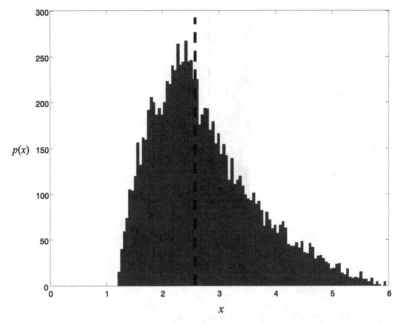

Fig. 7.22. Division of two uncorrelated ($\rho = 0$) uniform probability distributions. The dashed line shows the division of the means of the initial distributions.

Division

Let us now consider the division between the two uniform distributions P_1 and P_2 defined above.

Figure 7.22 shows the obtained histogram, when a Monte Carlo simulation (by means of 10,000 extractions) is performed, under the assumption of total uncorrelation of the two distributions. Figure 7.23 shows the histogram obtained from a Monte Carlo simulation under the assumption of total positive correlation between the two considered distributions, whereas Fig. 7.24 shows the histogram obtained from a Monte Carlo simulation under the assumption of total negative correlation.

Similar considerations can be done as in the case of the product operation. In fact, the obtained histograms show absence of symmetry and do not tend to the normal or uniform distribution. Also in this case, from an intuitive visual analysis, it could be assessed that the shape of the obtained histogram is similar to a chi-square distribution, when uncorrelated random variables are considered, and to an exponential distribution, when totally correlated random variables are considered. However, again, no analytical support is available in the literature to support this conclusion in general situations.

On the other hand, the theoretical supports, obtained by applying Eqs. (7.15), (7.16), and (7.17) in the given example, are confirmed by the

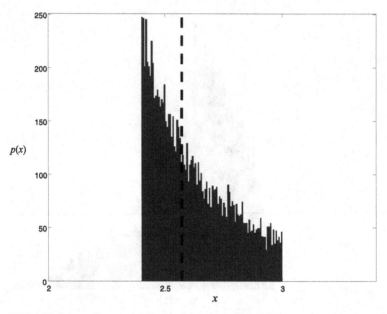

Fig. 7.23. Division of two totally positive correlated ($\rho = +1$) uniform probability distributions. The dashed line shows the division of the means of the initial distributions.

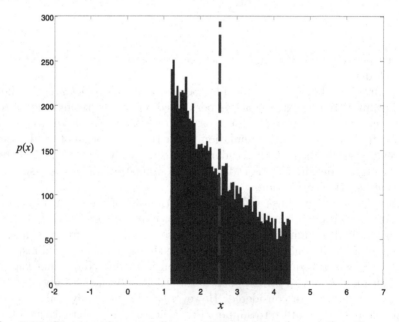

Fig. 7.24. Division of two totally negative correlated ($\rho = -1$) uniform probability distributions. The dashed line shows the division of the means of the initial distributions.

obtained results. In particular, the final support is $[1.2,6]$ when $\rho = 0$ or $\rho = -1$ is considered, and $[2.4,3]$ if $\rho = +1$ is considered.

7.2 Mathematics for the random parts of RFVs

The results obtained in the previous sections are the starting point for the definition of a suitable mathematics for the random part of an RFV. This mathematics should in fact be able to represent the behavior of random contributions when they combine with each other.

Unfortunately, the above case study led to strict results only when sum and difference were considered, whereas the more complicated cases of product and division do not still lead to a totally satisfying conclusion.

For these reasons, an accurate mathematics can be defined in the case of sums and differences between RFVs, whereas only an approximation can be given in the case of products and divisions.[7]

Let us consider, however, that the supports of the final distributions can be always accurately evaluated, and because the support of a distribution determines the confidence interval at level of confidence 1, this operation is surely the most important one.

Let A and B be two RFVs that only represent random contributions. In other words, A and B are RFVs for which the width of the internal membership functions is nil, and the external membership functions represent only the effects of random phenomena. An example is given in Fig. 7.25. Let us consider that RFVs A and B in this figure are the RFVs that represent, respectively, the probability distributions P_1 and P_2 defined in the previous section. In fact, if the probability–possibility transformation defined in Chapter 5 is applied, a uniform probability distribution is equivalent to a triangular possibility distribution.

Let us now consider the four arithmetic operations, for the different values of the correlation coefficient, and let us follow two separate steps. The first one is related to the determination of the support of the final RFV and the second one is related to the determination of the shape of its external membership function.

Therefore, the first step allows one to determine the α-cut at level $\alpha = 0$ or, in other words, the confidence interval at level of confidence 1 of the result. On the other hand, the second step allows one to determine the possibility

---

[7] This does not exclude, however, that this mathematics could be refined in the future and more accurate results could be obtained. To this respect, let us consider that the importance of the RFVs in the metrological field is already recognized, and new definitions can only lead to more accurate solutions for the final distribution of the random part of the RFVs.

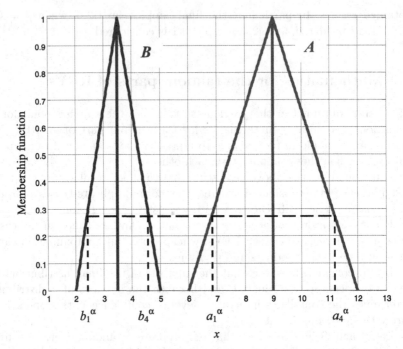

Fig. 7.25. Example of two RFVs with nil internal membership functions.

distribution of the result[8] and, therefore, all other confidence intervals (α-cuts).

As the confidence interval at level of confidence 1 is generally the most important information, it can be stated that the first step is the more important one.

7.2.1 Sum

Let us consider the sum $A + B$. Let $[a_1^\alpha, a_4^\alpha]$ and $[b_1^\alpha, b_4^\alpha]$ be the α-cuts, at the generic level α, of RFVs A and B, respectively.[9]

Let us first suppose that A and B are uncorrelated. When $\rho = 0$ is considered, the following steps must be followed.

---

[8] This applies because RFVs with nil internal membership functions are considered. In more general situations, this second step determines the possibility distribution of the random part of the RFV. In order to obtain the final possibility distribution (external membership function of the final RFV), it is necessary to consider also the internal membership function, as also shown in Chapter 5.

[9] As the internal interval of each α-cut degenerates into a single value, it can be not considered for the moment.

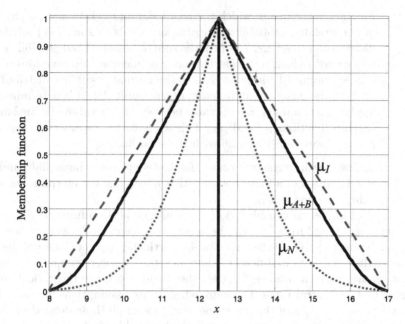

Fig. 7.26. Sum of RFVs A and B of Fig. 7.25 under the hypothesis of no correlation. The solid line is the final result. The dashed line is the membership function obtained by applying Eq. (7.18). The dotted line is the normal possibility distribution.

- Equation (7.1) is applied to all α-cuts of A and B at the same level α, that is,

$$[a_1^\alpha + b_1^\alpha, a_4^\alpha + b_4^\alpha] \tag{7.18}$$

This leads to the membership function μ_I (I stands for 'determined from intervals') reported in Fig. 7.26 with the dashed line.

- The normal possibility distribution[10] with mean value

$$m = \frac{(a_1^{\alpha=1} + b_1^{\alpha=1}) + (a_4^{\alpha=1} + b_4^{\alpha=1})}{2}$$

and standard deviation[11]

$$\sigma = \frac{1}{3} \min\{m - a_1^{\alpha=0} - b_1^{\alpha=0}; a_4^{\alpha=0} + b_4^{\alpha=0} - m\}$$

is determined. This leads to the membership function μ_N (N stands for 'normal') reported in Fig. 7.26 with the dotted line. This membership

---

[10] The term 'normal possibility distribution' refers to the membership function obtained by applying the probability–possibility transformation defined in Chapter 5 to a normal probability distribution.

[11] As a normal distribution is taken into account, the confidence interval at level of confidence 1 is supposed to be the $\pm 3\sigma$ interval around the mean value. See Chapter 5 for more details.

function is the one to which the result should tend. In fact, as the sum of uncorrelated probability distributions tends to a normal probability distribution (see Figs. 7.7–7.9), the sum of 'uncorrelated possibility distributions' (which represent the external membership functions of the corresponding RFVs) must tend to the normal possibility distribution.

- At last, the membership function of the result $A + B$ is determined by applying a suitable fuzzy operator between the two above defined membership functions μ_I and μ_N. This leads to the membership function μ_{A+B} reported in Fig. 7.26 with the solid line.

Of course, the implementation of this last step is not immediate and requires the definition of a fuzzy operator that allows one to determine the final membership function μ_{A+B}.

First of all, let us consider that the final membership function μ_{A+B} lies in between μ_N and μ_I, which can be considered, respectively, as the lower and upper bounds of the possible results. In particular, μ_I could be obtained if the results coming from different random extractions were supposed to never compensate with each other. On the other hand, μ_N is the theoretical bound given by the Central Limit Theorem, when all its assumptions are met. This restricts the selection of the fuzzy operator, among all those defined in Chapter 6, to the averaging operators. In fact, all possible kinds of t-norms would provide a membership function that is smaller than the smaller function between μ_N and μ_I; similarly, all possible kinds of t-conorms would provide a membership function that is greater than the greater function between μ_N and μ_I. On the contrary, only the averaging operators provide a membership function within μ_N and μ_I.

Among the various averaging operators defined in Chapter 6, the one that seems to be the most appropriate is an OWA operator. To fully understand this fact, let us first remember the definition of an OWA operator in the simplified case of $n = 2$, because only two aggregates μ_N and μ_I are present in our case.

For every value x belonging to the universal set, let us consider

$$a = \mu_N(x) \quad \text{and} \quad b = \mu_I(x)$$

then, the OWA operators define the membership grade c of the final membership function as

$$c = M_w(a, b) = w_1 \ \max\{a, b\} + w_2 \ \min\{a, b\} \qquad (7.19)$$

where $w_1 + w_2 = 1$, as shown in Fig. 7.27.

Let us remember that $\mathbf{w} = \langle w_1, w_2 \rangle$ is the *weighting vector*, and $w_i \in [0, 1]$. Let us also remember that the fundamental aspect of this operator is the reordering step. In fact, each aggregate (a or b) is not associated with a particular weight w_i, but rather a weight is associated with a particular ordered position of the aggregates. In other words, weight w_1 is always associated with

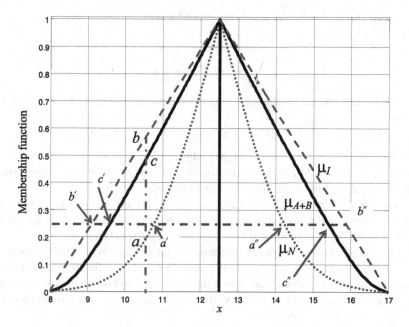

Fig. 7.27. Application of the OWA operator to the membership functions μ_I and μ_N. The dashed-dotted vertical line shows the vertical definition (in terms of membership grades); the dash-dotted line shows the horizontal definition (in terms of levels α).

the greater aggregate, and weight w_2 is always associated with the smaller aggregate.

Moreover, as proved in Chapter 6, in this particular case, the OWA operator can be defined on the α-cuts, instead of the membership grades, and provides an RFV as its result. This also allows one to have always the same computational burden, because the resolution on the α-levels can be a priori set, independently from the width of the universal set, or, equivalently, independently from the width of the supports of μ_I and μ_N.

In this respect, let us consider again Fig. 7.27. Then, two considerations can be done, as follows:

- The OWA operator must be applied twice for each considered level α.
- A reordering step still applies. In fact, although from a vertical point of view, the two weights w_1 and w_2 are applied, respectively, to the greater and smaller aggregate, when the 'horizontal' approach is followed, the two weights must be applied, respectively, to the external and internal aggregates.

From the first consideration, we should have

$$c' = w'_1 \, b' + w'_2 \, a'$$

$$c'' = w''_1 \, b'' + w''_2 \, a''$$

If also the second consideration is taken into account, then we generically have

$$c = M_w(a, b) = w_1 \, \text{ext}(a, b) + w_2 \, \text{int}(a, b) \tag{7.20}$$

where a and b are the left (or right) edges of the α-cuts of μ_I and μ_N, and $\text{ext}(a, b)$ and $\text{int}(a, b)$ are, respectively, the left (or right) edge of the α-cut of the external and internal membership function.

Equation (7.20) is applied twice: once on the left and once on the right edges of the α-cuts of μ_I and μ_N. In this way, the left and right edges of the α-cut of the final membership function (μ_{A+B}) are determined.

At this point, it is necessary to define weights w_1 and w_2. In this respect, let us consider again, only as the starting point, probability distribution functions and statistics. It is known that, when the mean of n equal uncorrelated random variables is taken, the standard deviation of the mean is $\sqrt{n}$ times smaller than the initial standard deviation.

Hence, if we consider that the OWA operator can be also seen as a weighted mean of μ_I and μ_N, then it can be stated that, for each considered level α, the width of the interval between μ_N and μ_{A+B} is $\sqrt{2}$ times smaller than the width of the interval between μ_I and μ_N; that is,

$$|c - \text{int}(a, b)| = \frac{|b - a|}{\sqrt{2}} = k \, |b - a| \tag{7.21}$$

where $k = 1/\sqrt{2}$.

Let us now consider, separately, the case where a and b are the right edges of the α-cuts of μ_I and μ_N, and the case where a and b are the left edges. In the first case, Eq. (7.21) can be rewritten as (see Fig. 7.27):

$$c - a = k \, (b - a)$$

and

$$c = a + k \, (b - a) = a \, (1 - k) + b \, k \tag{7.22}$$

In the second case, Eq. (7.21) can be rewritten as (see Fig. 7.27)

$$a - c = k \, (a - b)$$

which again leads to Eq. (7.22).

By considering Eq. (7.22), Eq. (7.20) can be rewritten as

$$c = k \, \text{ext}(a, b) + (1 - k) \, \text{int}(a, b) \tag{7.23}$$

Let us underline that, because of the ordering step (provided by functions ext and int), because the weights (k and $1-k$, $k = 1/\sqrt{2}$) are in the range [0,1] and because the sum of the weights is one, the fuzzy operator (7.23) classifies as an OWA operator. It can be readily proven that Eq. (7.23) satisfies to the properties of symmetry, idempotency, monotonicity, continuity, and the boundary conditions (6.30).

By applying Eq. (7.23), the membership function μ_{A+B} reported in Fig. 7.26 with the solid line can be obtained. This means that, after applying the averaging operator between μ_I and μ_N, the result μ_{A+B} represents an approximation of the normal possibility distribution. Moreover, the more combinations are performed, the better is the attained approximation.

Let us now suppose that A and B are totally correlated.

When $\rho = +1$ is considered, that is, a total positive correlation, each α-cut of the result is obtained according to Eq. (7.18) as

$$[a_1^\alpha + b_1^\alpha, a_4^\alpha + b_4^\alpha]$$

without any need of further elaborations.

In fact, according to the way random effects combine together, the confidence intervals of the final result, for all possible levels of confidence, are obtained from the corresponding confidence intervals of the initial distributions in the same way as the support (which is the confidence interval with level of confidence 1), as shown in Fig. 7.3. This is also consistent with the example of Fig. 7.10.

The application of the above equation to RFVs A and B shown in Fig. 7.25 leads to the result shown in Fig. 7.28. It can be noted that the shape of the RFV in Fig. 7.28 is triangular, as well as those of the initial RFVs, as expected from the above considerations.

When $\rho = -1$ is considered, that is a total negative correlation, Eq. (7.2) could be applied to each α-cut. It would lead to

$$[\min\{a_1^\alpha + b_4^\alpha; a_4^\alpha + b_1^\alpha\}, \max\{a_1^\alpha + b_4^\alpha; a_4^\alpha + b_1^\alpha\}] \qquad (7.24)$$

However, in this case, if the two external membership functions of A and B do not have the same shape, Eq. (7.24) could provide a multivalued function, that does not satisfy the requirements of a membership function.

The analysis, reported in the previous section, of the behaviour of random contributions when they combine together shows that the distributions obtained under total positive ($\rho = +1$) and total negative ($\rho = -1$) correlation have the same shape, but different supports, as shown in Figs. 7.10 and 7.13.

Therefore, under total negative correlation, each α-cut can be determined starting again from (7.18) and scaling it according to the ratio between the supports obtained for ρ=-1 and ρ=+1. The scaling factor s is hence given by:

$$s = \frac{\max\left\{a_1^{\alpha=0} + b_4^{\alpha=0}; a_4^{\alpha=0} + b_1^{\alpha=0}\right\} - \min\left\{a_1^{\alpha=0} + b_4^{\alpha=0}; a_4^{\alpha=0} + b_1^{\alpha=0}\right\}}{(a_4^{\alpha=0} + b_4^{\alpha=0}) - (a_1^{\alpha=0} + b_1^{\alpha=0})}$$

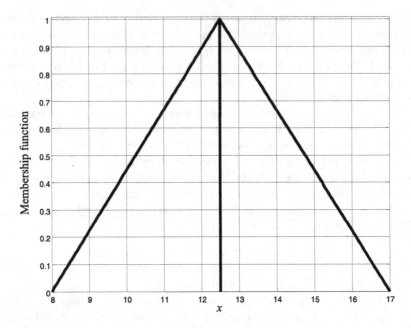

Fig. 7.28. Sum of RFVs A and B of Fig. 7.25 under the hypothesis of total positive correlation.

Then the result shown in Fig. 7.29 is obtained, if the RFVs shown in Fig. 7.25 are considered. The result shows again a triangular membership function, as expected.

When a different value for the correlation coefficient ρ is considered, intermediate results can be expected. In particular, a correlation coefficient ρ in the range $(0, 1)$ is expected to provide a result, whose membership function is between the result obtained for $\rho = 0$ and that obtained for $\rho = 1$. On the other hand, a correlation coefficient ρ in the range $(-1, 0)$ is expected to provide a result, whose membership function is between the result obtained for $\rho = 0$ and that obtained for $\rho = -1$. This is shown, for the same RFVs shown in Fig. 7.25, in Fig. 7.30.

Hence, to obtain the desired result, an OWA operator can be applied again.

For every level α, let us call c, b_+ and b_- the right edges of the α-cuts of the membership functions obtained for $\rho = 0$, $+1$ and -1, respectively, as shown in Fig. 7.30. Let us also call d_+ and d_- the desired results for a positive and a negative value of ρ, respectively.

Then, the following OWA operators can be defined:

$$d_+ = d_+(c, b_+) = (1 - \rho)\, c + \rho\, b_+ \tag{7.25}$$

$$d_- = d_-(c, b_-) = (1 + \rho)\, c - \rho\, b_- \tag{7.26}$$

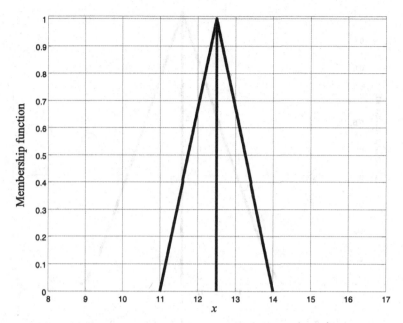

Fig. 7.29. Sum of RFVs A and B of Fig. 7.25 under the hypothesis of total negative correlation.

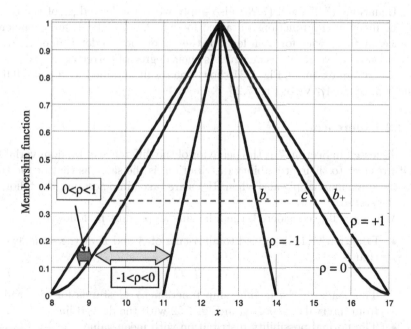

Fig. 7.30. Sum of RFVs A and B of Fig. 7.25 for a generic correlation coefficient ρ. The final membership function lies within one of the regions shown by the arrows.

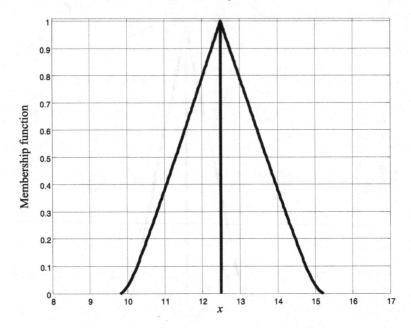

Fig. 7.31. Sum of RFVs A and B of Fig. 7.25 for $\rho = 0.6$.

Equations (7.25) and (7.26) also apply when the left edges of the α-cuts of the membership functions obtained for $\rho = 0, +1$ and -1, are considered. Hence, applied twice for each level α, they allow one to determine the result of the sum of two RFVs, whichever is their degree of correlation.

To give an example, Fig. 7.31 shows the result obtained with $\rho = -0.6$ for the sum of the RFVs of Fig. 7.25.

7.2.2 Difference

As discussed in Section 7.1, the behavior of the difference of random variables, with respect to their probability distribution, is the same as that of the sum. As the same must apply also when RFVs are taken into account, the following can be stated.

When $\rho = 0$ is considered, the following steps must be followed:

- Equation (7.4) is applied to all α-cuts of A and B at the same level α; that is,

$$[a_1^\alpha - b_4^\alpha, a_4^\alpha - b_1^\alpha] \tag{7.27}$$

  This leads to the membership function μ_I (I stands for 'determined from intervals') reported in Fig. 7.32 with the dashed line.
- The normal possibility distribution with mean value

$$m = \frac{(a_1^{\alpha=1} - b_4^{\alpha=1}) + (a_4^{\alpha=1} - b_1^{\alpha=1})}{2}$$

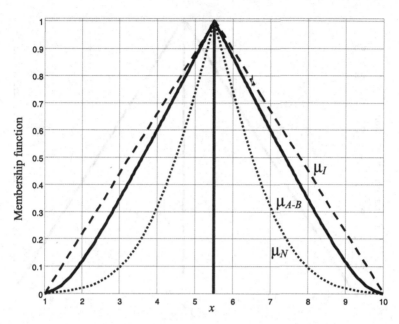

Fig. 7.32. Difference of RFVs A and B of Fig. 7.25 under the hypothesis of no correlation. The solid line is the final result. The dashed line is the membership function obtained by applying Eq. (7.27). The dotted line is the normal possibility distribution.

and standard deviation

$$\sigma = \frac{1}{3} \min\{m - a_1^{\alpha=0} + b_4^{\alpha=0}; a_4^{\alpha=0} - b_1^{\alpha=0} - m\}$$

is determined. This leads to the membership function μ_N (N stands for 'normal') reported in Fig. 7.32 with the dotted line. This membership function is the one to which the result should tend. In fact, as the difference of uncorrelated probability distributions tends to a normal probability distribution, the difference of 'uncorrelated possibility distributions' (which represent the external membership functions of the corresponding RFVs) must tend to the normal possibility distribution.

- At last, the membership function of the result $A - B$ is determined by applying the same OWA operator (7.23) to the two above-defined membership functions μ_I and μ_N. This leads to the membership function μ_{A-B} reported in Fig. 7.32 with the solid line. It can be noted that, with respect to μ_I, the result represents an approximation of the normal possibility distribution, as also expected by considering Fig. 7.16.

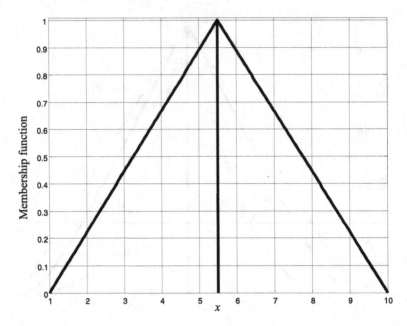

Fig. 7.33. Difference of RFVs A and B of Fig. 7.25 under the hypothesis of total negative correlation.

When $\rho = -1$ is considered, each α-cut of the result is obtained according to Eq. (7.27). If the two RFVs A and B of Fig. 7.25 are considered again, this leads to the result shown in Fig. 7.33.

When $\rho=+1$ is considered, the same considerations apply, as those reported for the sum in the case of total negative correlation. In fact, if Eq. (7.5) is applied to all α-cuts of A and B, it provides:

$$[\min\{a_1^\alpha - b_1^\alpha; a_4^\alpha - b_4^\alpha\}, \max\{a_1^\alpha - b_1^\alpha; a_4^\alpha - b_4^\alpha\}] \tag{7.28}$$

that, if the two external membership functions of A and B do not have the same shape, could provide a multivalued function, and hence cannot be employed to represent a membership function.

Similarly to the case of the sum, the distributions obtained under total negative ($\rho = -1$) and total positive ($\rho = +1$) correlation have the same shape, but different supports (see Figs 7.17 and 7.18).

Therefore, under total positive correlation, each α-cut can be determined starting again from (7.27) and scaling it according to the ratio between the supports obtained for $\rho = +1$ and $\rho = -1$. The scaling factor s is hence given by:

$$s = \frac{\max\left\{a_1^{\alpha=0} - b_1^{\alpha=0}; a_4^{\alpha=0} - b_4^{\alpha=0}\right\} - \min\left\{a_1^{\alpha=0} - b_1^{\alpha=0}; a_4^{\alpha=0} - b_4^{\alpha=0}\right\}}{(a_4^{\alpha=0} - b_1^{\alpha=0}) - (a_1^{\alpha=0} - b_4^{\alpha=0})}$$

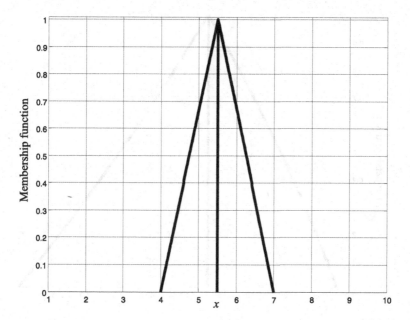

Fig. 7.34. Difference of RFVs A and B of Fig. 7.25 under the hypothesis of total positive correlation.

When the two RFVs A and B of Fig. 7.25 are considered, this leads to the result shown in Fig. 7.34.

When a different value for the correlation coefficient ρ is considered, intermediate results can be expected. Similar considerations as those made for the sum can be drawn again. Therefore, the OWA operators (7.25) and (7.26) must be applied to determine the actual result.

7.2.3 Product

Let us now consider the product $A * B$, where A and B are, again, the RFVs shown in Fig. 7.25. When $\rho = 0$ is considered, Eq. (7.6) is applied to all α-cuts of A and B at the same level α; that is,

$$[\min\{a_1^\alpha\, b_1^\alpha; a_1^\alpha\, b_4^\alpha; a_4^\alpha\, b_1^\alpha; a_4^\alpha\, b_4^\alpha\}, \max\{a_1^\alpha\, b_1^\alpha; a_1^\alpha\, b_4^\alpha; a_4^\alpha\, b_1^\alpha; a_4^\alpha\, b_4^\alpha\}] \quad (7.29)$$

It can be readily proven that Eq. (7.29) provides a membership function [KY95]–[KG91]. In this case, the result is shown in Fig. 7.35.

When $\rho = +1$ is considered, Eq. (7.7) could be applied to all α-cuts. Let us then call c_1^α and c_4^α the left and right edges, respectively, of the final result. Hence, called

$$x_m = \frac{\mu_1 - \mu_2\, r}{2}$$

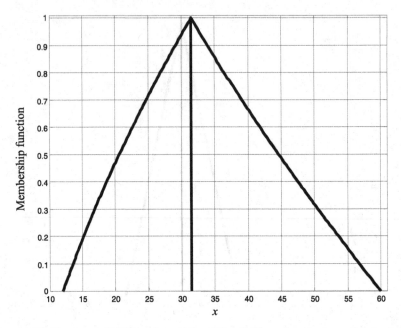

Fig. 7.35. Product of RFVs A and B of Fig. 7.25 under the hypothesis of total uncorrelation. The same result is also obtained under the hypothesis of total positive correlation.

and

$$y_m = \mu_2 + \frac{x_m - \mu_1}{r}$$

where

$$\mu_1 = \frac{a_1^\alpha + a_4^\alpha}{2}$$

$$\mu_2 = \frac{b_1^\alpha + b_4^\alpha}{2}$$

$$r = \frac{a_4^\alpha - a_1^\alpha}{b_4^\alpha - b_1^\alpha}$$

are, respectively, the mean value of the α-cut $[a_1^\alpha, a_4^\alpha]$, the mean value of the α-cut $[b_1^\alpha, b_4^\alpha]$, and the ratio between their widths, it is

$$c_1^\alpha = \begin{cases} x_m\, y_m & \text{if } a_1^\alpha < x_m < a_4^\alpha \\ \min\{a_1^\alpha b_1^\alpha, a_4^\alpha b_4^\alpha\} & \text{otherwise} \end{cases} \tag{7.30}$$

$$c_4^\alpha = \max\{a_1^\alpha b_1^\alpha, a_4^\alpha b_4^\alpha\}$$

However, this result could be incorrect, since Eq. (7.30) could lead to a multivalued function, that does not satisfy the requirements of a membership function. In fact, the α-cuts at levels $\alpha > 0$, that represent confidence intervals at levels of confidence $1 - \alpha$, could be wider than the α-cut at level $\alpha = 0$ that, being the support of the final RFV, is the interval of confidence at level of confidence 1. This condition is recognized if at least one couple of values α_1 and $\alpha_2, \alpha_1 < \alpha_2$, exists so that

$$c_1^{\alpha_1} > c_1^{\alpha_2} \text{ or } c_4^{\alpha_1} < c_4^{\alpha_2}$$

Since this is physically not allowable, the membership function provided by Eq. (7.30) must be, when necessary, suitably adjusted to force all α-cuts to be nested. This can be obtained by forcing:

$$c_1^{\alpha_2} = c_1^{\alpha_1}$$

for all values α_1 and $\alpha_2, \alpha_1 < \alpha_2$, for which $c_1^{\alpha_2} < c_1^{\alpha_1}$ and

$$c_4^{\alpha_2} = c_4^{\alpha_1}$$

for all values α_1 and $\alpha_2, \alpha_1 < \alpha_2$, for which $c_4^{\alpha_2} > c_4^{\alpha_1}$.

For the RFVs shown in Fig. 7.25, this correction is not necessary. However, Eq. (7.30) leads again to the RFV shown in Fig. 7.35, that is the RFV obtained under total uncorrelation.

This result needs some further considerations. In fact, if Figs. 7.19 and 7.20 are considered again, obtained by applying the Monte Carlo simulation to the probability distribution functions from which A and B are derived, they show two different probability distributions, distributed over the same interval, which is also the same interval considered in Fig. 7.35.

If the two probability distributions of Figs. 7.19 and 7.20 were transformed into possibility distributions, as described in Chapter 5, two different results would be, of course, obtained. To obtain two different RFVs for the two considered cases of zero and unitary correlation coefficients, an OWA operator should be applied after Eq. (7.29), when $\rho = 0$, similarly to what is done for the sum and the difference operations. However, when sum and difference are considered, the probability distribution to which the result tends is well known, whereas the same cannot be said for the product operation, and neither for the division operation, which will be considered in next section.

If the probability distribution to which the result should tend is not known, it is also not known the possibility distribution to which the result should tend. For this reason, it is not possible to apply an OWA operator, because the second aggregate is not known.

Nevertheless, in this way, the confidence interval at level of confidence 1 and the confidence interval at level of confidence 0 are correctly determined. On the other hand, all intermediate confidence intervals are overestimated.

Let us consider, in this respect, that overestimation is always preferable, as far as measurement uncertainty is concerned and that, if an OWA operator were considered among wrong aggregates, this could lead to an intolerable

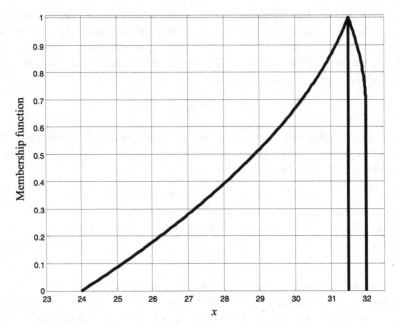

Fig. 7.36. Product of RFVs A and B of Fig. 7.25 under the hypothesis of total negative correlation.

underestimation. Therefore, when $\rho = 0$ and the product is taken into account, the implementation of Eq. (7.29), without any further step,[12] seems to be the best solution, because it is the most conservative one.

When $\rho = -1$ is considered, Eq. (7.11) could be applied to all α-cuts; that is,

$$c_1^\alpha = \min\{a_1^\alpha b_4^\alpha, a_4^\alpha b_1^\alpha\}$$

$$c_4^\alpha = \begin{cases} x_M \, y_M & \text{if } a_1^\alpha < x_M < a_4^\alpha \\ \max\{a_1^\alpha b_4^\alpha, a_4^\alpha b_1^\alpha\} & \text{otherwise} \end{cases} \qquad (7.31)$$

where

$$x_M = \frac{\mu_1 + \mu_2 \, r}{2}$$

and

$$y_M = \mu_2 - \frac{x_M - \mu_1}{r}$$

Similarly to the case of total positive correlation, Eq. (7.31) could lead to a multivalued function. This condition is recognized in the same way as for

---

[12] This does not exclude, however, that this mathematics could be refined in the future and more accurate results could be obtained.

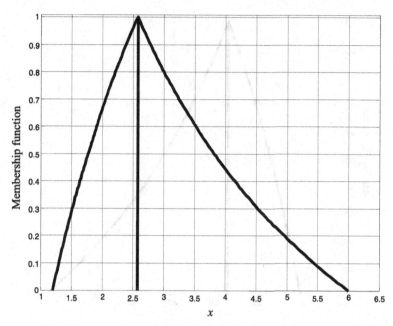

Fig. 7.37. Division of RFVs A and B of Fig. 7.25 under the hypothesis of no correlation. The same result is also obtained under the hypothesis of total negative correlation.

the case $\rho = +1$ and requires the same correction. For the RFVs shown in Fig. 7.25, this is not necessary and the result is shown in Fig. 7.36.

When a different value for the correlation coefficient is considered, intermediate results can be expected. Similar considerations as those made for the sum can be drawn again. Therefore, the OWA operators (7.25) and (7.26) must be applied to determine the actual result.

7.2.4 Division

Let us now consider the division A/B, where A and B are, again, the RFVs shown in Fig. 7.25.

When $\rho = 0$ is considered, Eq. (7.15) is applied to all α-cuts of A and B at the same level α; that is,

$$[\min\{a_1^\alpha/b_1^\alpha; a_1^\alpha/b_4^\alpha; a_4^\alpha/b_1^\alpha; a_4^\alpha/b_4^\alpha\}, \max\{a_1^\alpha/b_1^\alpha; a_1^\alpha/b_4^\alpha; a_4^\alpha/b_1^\alpha; a_4^\alpha/b_4^\alpha\}] \quad (7.32)$$

It can be readily proven that Eq. (7.32) provides a membership function [KY95]–[KG91]. In this case, the result is shown in Fig. 7.37.

When $\rho = +1$ is considered, Eq. (7.16) is applied to all α-cuts of A and B at the same level α; that is,

$$[\min\{a_1^\alpha/b_1^\alpha; a_4^\alpha/b_4^\alpha\}, \max\{a_1^\alpha/b_1^\alpha; a_4^\alpha/b_4^\alpha\}] \quad (7.33)$$

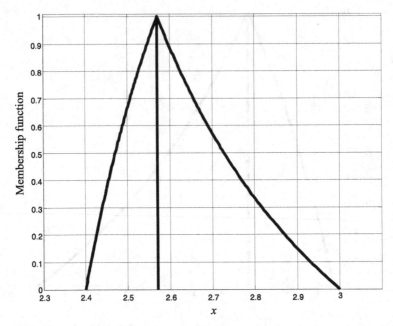

Fig. 7.38. Division of RFVs A and B of Fig. 7.25 under the hypothesis of total positive correlation.

The same considerations as those reported in section 7.2.3 still apply, and the same correction as the one given in that section must be applied, if necessary, to the function provided by Eq. (7.33) to obtain a membership function.

When $\rho = +1$, the division A/B, where A and B are the RFVs shown in Fig. 7.25, leads to the result shown in Fig. 7.38.

When $\rho = -1$ is considered, Eq. (7.17) is applied to all α-cuts of A and B at the same level α; that is,

$$[\min\{a_1^\alpha/b_4^\alpha; a_4^\alpha/b_1^\alpha\}, \max\{a_1^\alpha/b_4^\alpha; a_4^\alpha/b_1^\alpha\}] \tag{7.34}$$

Again, the same considerations and the same corrections as those reported in section 7.2.3 apply.

For the two considered RFVs, this leads again to the result shown in Fig. 7.37.

Similar considerations as those done for the case of the product can be done here. In fact, if Figs. 7.22 and 7.24 are considered, it follows that the two probability distributions have the same support, but different distribution functions, whereas Eqs. (7.32) and (7.34) lead, in this particular example, to the same RFV.

However, this is actually the best result for at least two reasons. First of all, the application of the above-defined equations lead to a correct estimation of the confidence intervals at levels of confidence 0 and 1 and to an overestimation of the confidence intervals for all levels of confidence between 0 and 1. Moreover, no OWA operator can be correctly applied unless the possibility distribution function to which the result should tend is known.

When a different value for the correlation coefficient ρ is considered, intermediate results can be expected. Similar considerations as those made for the sum can be drawn again. Therefore, the OWA operators (7.25) and (7.26) must be applied to determine the actual result.

7.3 The complete mathematics

In the previous section, the mathematics of the random part of RFVs has been defined, in the simplified case where the internal membership functions of the considered RFVs had null width. This case was fully exhaustive, being the internal membership function related to nonrandom phenomena.

However, in the more general case, when also nonrandom phenomena must be taken into account, a more complete mathematics must be employed.

Let us consider two RFVs A and B, and let $[a_1^\alpha, a_2^\alpha, a_3^\alpha, a_4^\alpha]$ and $[b_1^\alpha, b_2^\alpha, b_3^\alpha, b_4^\alpha]$ be their generic α-cuts. Different situations must be considered.

The case where only random contributions are present in both RFVs; that is, for every α in the range $[0, 1]$,

$$a_2^\alpha = a_3^\alpha \quad \text{and} \quad b_2^\alpha = b_3^\alpha$$

has already been discussed in the previous section.

The case where one or both RFVs show only nonrandom contributions; that is, for every α in the range $[0, 1]$,

$$a_1^\alpha = a_2^\alpha \qquad\qquad b_1^\alpha = b_2^\alpha$$

$$\text{and/or}$$

$$a_3^\alpha = a_4^\alpha \qquad\qquad b_3^\alpha = b_4^\alpha$$

is discussed in Section 7.3.1.

When both the random and the systematic parts are present in both considered RFVs, the two contributions must be properly taken into account, by composing the same kind of contributions together. However, it is again possible to distinguish among two situations, as shown in Figs. 7.39 and 7.40, respectively.

The RFVs in Fig. 7.39 have the following properties:

$$a_2^\alpha = a_2^{\alpha'} \qquad\qquad b_2^\alpha = b_2^{\alpha'}$$

$$\text{and}$$

$$a_3^\alpha = a_3^{\alpha'} \qquad\qquad b_3^\alpha = b_3^{\alpha'}$$

for every α and α' in the range $[0,1]$. In other words, the internal membership function of both RFVs is rectangular.

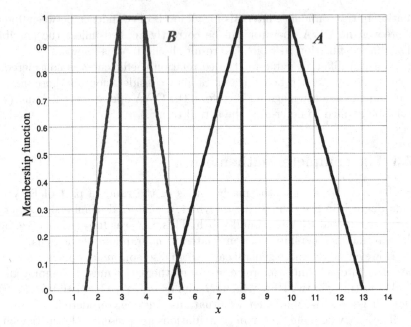

Fig. 7.39. Example of two RFVs.

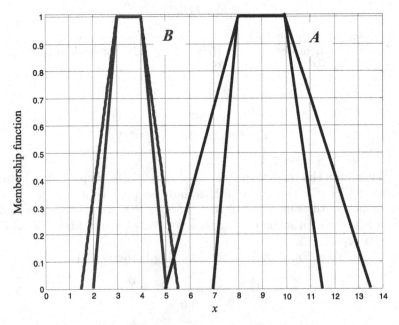

Fig. 7.40. Example of two generic RFVs.

As discussed in the previous chapters, this case is the most common one. In fact, when an RFV is used to represent a measurement result, generally both random and nonrandom contributions affect the measurement result itself. Moreover, the nonrandom contributions are generally given in terms of a confidence interval, within which the results that could be reasonably attributed to the measurand are supposed to lie; if no additional information is available, as in most practical cases, this situation is described, as widely discussed in Chapters 2 and 3, by a rectangular possibility distribution, that is, by a rectangular fuzzy variable. This situation is thoroughly discussed in Section 7.3.2.

At last, Section 7.3.3 briefly discusses the more general situation, where both RFVs A and B present both random and nonrandom contributions, and the shape of the internal membership function is no longer rectangular, as shown in the example of Fig. 7.40.

In the following sections, $[c_1^\alpha, c_2^\alpha, c_3^\alpha, c_4^\alpha]$ is used to denote the generic α-cut of the result.

As all explanations about the behavior of both the random and the nonrandom parts of the RFVs when they compose with each other have been already given (in this chapter and in Chapter 2), the formulas are reported below, without any further discussion.

7.3.1 One or both RFVs show only nonrandom contributions

When an RFV represents a measurement result affected only by systematic or unknown contributions, the internal and external membership functions coincide. It can be also stated that the RFV degenerates into a pure fuzzy variable or, in more mathematical terms, into a fuzzy variable of type 1.

If two RFVs of this kind, as shown, for instance, in Fig. 7.41, are composed together, the internal and external membership functions of the result must necessarily coincide.

As fuzzy variables of type 1 are composed according to the mathematics of the intervals, as shown in Chapter 2, it follows that, when both the considered RFVs A and B degenerate into fuzzy variables of type 1, this mathematics must be applied to both the internal and the external membership functions.

This applies also when only one of the two considered RFVs degenerates into a fuzzy variable of type 1, whereas the other one presents also (or only) effects due to random contributions to uncertainty. An example is shown in Fig. 7.42.

In fact, in this particular situation, only one of the two considered measurement results is affected by random effects. In this case, in general, no

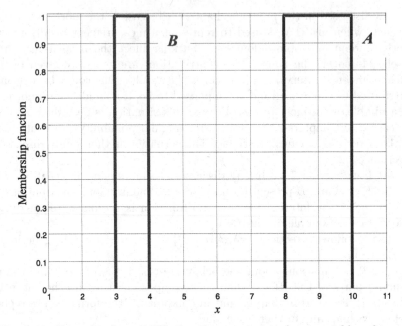

Fig. 7.41. Example of two RFVs that degenerate into fuzzy variables of type 1.

compensation can be assumed to occur during the composition.[13] For the sake of completeness, the final mathematics is reported here.

Sum

$$c_1^\alpha = a_1^\alpha + b_1^\alpha$$
$$c_2^\alpha = a_2^\alpha + b_2^\alpha$$
$$c_3^\alpha = a_3^\alpha + b_3^\alpha$$
$$c_4^\alpha = a_4^\alpha + b_4^\alpha$$

---

[13] Of course, it cannot be excluded, a priori, that additional evidence becomes available, in some particular cases, showing that some compensation might occur. This situation, however, should be treated in the RFV definition stage, by suitably assigning the internal and the external membership functions of the two RFVs.

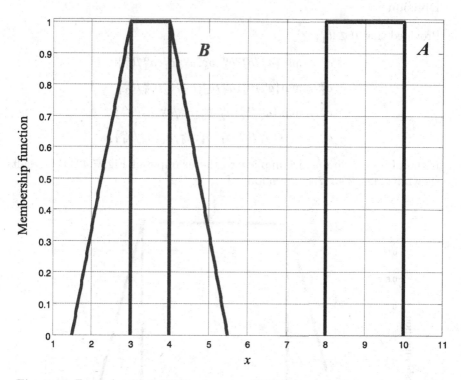

Fig. 7.42. Example of two RFVs, where one RFV degenerates into a fuzzy variable of type 1.

Difference

$$c_1^\alpha = a_1^\alpha - b_4^\alpha$$
$$c_2^\alpha = a_2^\alpha - b_3^\alpha$$
$$c_3^\alpha = a_3^\alpha - b_2^\alpha$$
$$c_4^\alpha = a_4^\alpha - b_1^\alpha$$

Product

$$c_1^\alpha = \min\{a_1^\alpha\, b_1^\alpha; a_1^\alpha\, b_4^\alpha; a_4^\alpha\, b_1^\alpha; a_4^\alpha\, b_4^\alpha\}$$
$$c_2^\alpha = \min\{a_2^\alpha\, b_2^\alpha; a_2^\alpha\, b_3^\alpha; a_3^\alpha\, b_2^\alpha; a_3^\alpha\, b_3^\alpha\}$$
$$c_3^\alpha = \max\{a_2^\alpha\, b_2^\alpha; a_2^\alpha\, b_3^\alpha; a_3^\alpha\, b_2^\alpha; a_3^\alpha\, b_3^\alpha\}$$
$$c_4^\alpha = \max\{a_1^\alpha\, b_1^\alpha; a_1^\alpha\, b_4^\alpha; a_4^\alpha\, b_1^\alpha; a_4^\alpha\, b_4^\alpha\}$$

Division

Provided that $0 \notin [b_1^\alpha; b_4^\alpha]$

$$c_1^\alpha = \min\{a_1^\alpha/b_1^\alpha; a_1^\alpha/b_4^\alpha; a_4^\alpha/b_1^\alpha; a_4^\alpha/b_4^\alpha\}$$
$$c_2^\alpha = \min\{a_2^\alpha/b_2^\alpha; a_2^\alpha/b_3^\alpha; a_3^\alpha/b_2^\alpha; a_3^\alpha/b_3^\alpha\}$$
$$c_3^\alpha = \max\{a_2^\alpha/b_2^\alpha; a_2^\alpha/b_3^\alpha; a_3^\alpha/b_2^\alpha; a_3^\alpha/b_3^\alpha\}$$
$$c_4^\alpha = \max\{a_1^\alpha/b_1^\alpha; a_1^\alpha/b_4^\alpha; a_4^\alpha/b_1^\alpha; a_4^\alpha/b_4^\alpha\}$$

If the above equations are applied to the example of Fig. 7.42, the results shown in Figs. 7.43–7.46 are obtained.

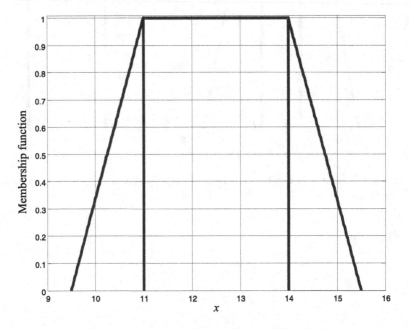

Fig. 7.43. Sum of the two RFVs of Fig. 7.42.

7.3.2 The internal membership function of both RFVs is rectangular

As discussed, in most measurement processes, all unknown and systematic effects are given in terms of a confidence interval, within which the possible measurement results are supposed to lie. As this situation is suitably described by a rectangular membership function, it follows that, in most practical applications, the measurement results can be described by RFVs like the ones in Fig. 7.39, for which the internal membership function is rectangular. In this

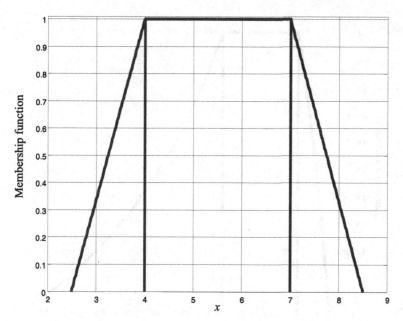

Fig. 7.44. Difference of the two RFVs of Fig. 7.42.

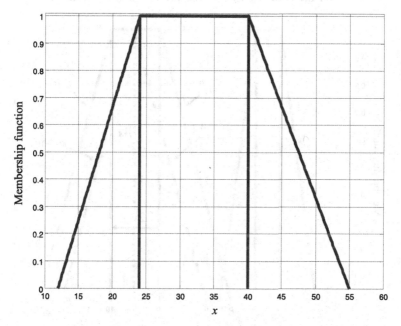

Fig. 7.45. Product of the two RFVs of Fig. 7.42.

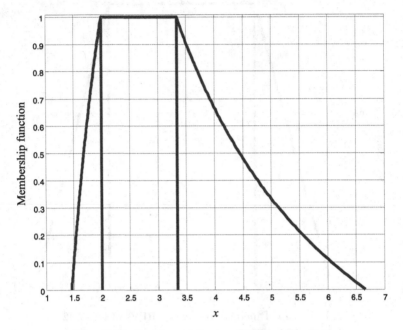

Fig. 7.46. Division of the two RFVs of Fig. 7.42.

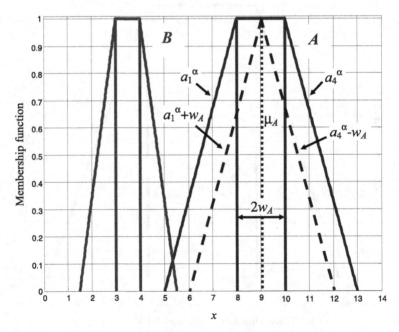

Fig. 7.47. Determination of the pure random part.

situation, the mathematics that must be applied to the RFVs is simpler than the most generic one, which should be applied in case the measurement results are represented by generic RFVs like the ones in Fig. 7.40. This is because the α-cuts of the internal fuzzy variable are all equals.

Let us consider that this also applies in the case in which the considered RFVs show only the random part, like the ones in Fig. 7.25. In fact, under this assumption, the internal fuzzy variable degenerates into a single scalar value, which, however, can be considered as the limit case of a rectangular fuzzy variable with zero width. Therefore, also this case can be treated by applying the mathematics given in this section.

When two RFVs have to be composed together, the same kind of contributions compose together obeying to their own composition rules. This requires one to separate the random part of the RFVs from the internal part, to process the two parts in a different way. In this respect, let us consider Fig. 7.47, which shows how the pure random part can be determined starting from the given RFV.

By definition, the mean value of an RFV A is determined as the mean value of the α-cut at level $\alpha = 1$; that is,[14]

$$\mu_A = \frac{a_2^{\alpha=1} + a_3^{\alpha=1}}{2}$$

Under the hypothesis of rectangular internal membership function, the semi-width of this last one is

$$w_A = \frac{a_3^{\alpha=1} - a_2^{\alpha=1}}{2}$$

Hence, the pure random part of the given RFV is a fuzzy variable (of type 1) whose α-cuts are

$$[a_1^\alpha + w_A; a_4^\alpha - w_A]$$

The following applies.

Sum

Let w_A and w_B be the semi-widths of the internal fuzzy variables of A and B, respectively:

$$w_A = \frac{a_3^{\alpha=1} - a_2^{\alpha=1}}{2}$$

$$w_B = \frac{b_3^{\alpha=1} - b_2^{\alpha=1}}{2}$$

---

[14] Let us remember that, by definition of an RFV, it is always

$$a_1^{\alpha=1} = a_2^{\alpha=1} \quad \text{and} \quad a_3^{\alpha=1} = a_4^{\alpha=1}$$

See Chapter 4 for more details.

Then, let us consider the following:

$$a_{r1}^\alpha = a_1^\alpha + w_A$$
$$a_{r4}^\alpha = a_4^\alpha - w_A$$
$$b_{r1}^\alpha = b_1^\alpha + w_B$$
$$b_{r4}^\alpha = b_4^\alpha - w_B$$

which represent the part of each α-cut associated with the pure random contributions.

Then,

- $\rho = 0$

$$c_1^\alpha(\rho = 0) = c_2^\alpha - \mu_c + k \cdot \text{ext}(a_{r1}^\alpha + b_{r1}^\alpha, g_1^\alpha) + (1 - k)$$
$$\cdot \text{int}(a_{r1}^\alpha + b_{r1}^\alpha, g_1^\alpha)$$
$$c_2^\alpha(\rho = 0) = a_2^\alpha + b_2^\alpha$$
$$c_3^\alpha(\rho = 0) = a_3^\alpha + b_3^\alpha$$
$$c_4^\alpha(\rho = 0) = c_3^\alpha - \mu_c + k \cdot \text{ext}(a_{r4}^\alpha + b_{r4}^\alpha, g_4^\alpha) + (1 - k)$$
$$\cdot \text{int}(a_{r4}^\alpha + b_{r4}^\alpha, g_4^\alpha)$$

where μ_c is the mean value of the sum:

$$\mu_c = \frac{c_2^{\alpha=1} + c_3^{\alpha=1}}{2}$$

k is a constant:

$$k = \frac{1}{\sqrt{2}}$$

and $[g_1^\alpha, g_4^\alpha]$ is the generic α-cut of the normal possibility distribution, having a mean value μ_c and standard deviation:

$$\sigma = \frac{1}{3} \min\{\mu_c - a_{r1}^{\alpha=0} - b_{r1}^{\alpha=0};\ a_{r4}^{\alpha=0} + b_{r4}^{\alpha=0} - \mu_c\},$$

- $\rho = +1$

$$c_1^\alpha(\rho = 1) = a_1^\alpha + b_1^\alpha$$
$$c_2^\alpha(\rho = 1) = a_2^\alpha + b_2^\alpha$$
$$c_3^\alpha(\rho = 1) = a_3^\alpha + b_3^\alpha$$
$$c_4^\alpha(\rho = 1) = a_4^\alpha + b_4^\alpha$$

- $\rho = -1$

$$c_1^\alpha(\rho = -1) = c_2^\alpha + s_1 \cdot (a_{r1}^\alpha + b_{r1}^\alpha - \mu_c)$$

$$c_2^\alpha(\rho = -1) = a_2^\alpha + b_2^\alpha$$

$$c_3^\alpha(\rho = -1) = a_3^\alpha + b_3^\alpha$$

$$c_4^\alpha(\rho = -1) = c_3^\alpha + s_4 \cdot (a_{r4}^\alpha + b_{r4}^\alpha - \mu_c)$$

where μ_c is the mean value of the sum,

$$s_1 = \frac{\mu_c - \min\left\{a_{r1}^{\alpha=0} + b_{r4}^{\alpha=0}; a_{r4}^{\alpha=0} + b_{r1}^{\alpha=0}\right\}}{\mu_c - (a_{r1}^{\alpha=0} + b_{r1}^{\alpha=0})}$$

and

$$s_4 = \frac{\max\left\{a_{r1}^{\alpha=0} + b_{r4}^{\alpha=0}; a_{r4}^{\alpha=0} + b_{r1}^{\alpha=0}\right\} - \mu_c}{(a_{r4}^{\alpha=0} + b_{r4}^{\alpha=0}) - \mu_c}$$

- $0 < \rho < 1$

$$c_1^\alpha(\rho > 0) = (1 - \rho)\, c_1^\alpha(\rho = 0) + \rho\, c_1^\alpha(\rho = 1)$$

$$c_2^\alpha(\rho > 0) = a_2^\alpha + b_2^\alpha$$

$$c_3^\alpha(\rho > 0) = a_3^\alpha + b_3^\alpha$$

$$c_4^\alpha(\rho > 0) = (1 - \rho)\, c_4^\alpha(\rho = 0) + \rho\, c_4^\alpha(\rho = 1)$$

- $-1 < \rho < 0$

$$c_1^\alpha(\rho < 0) = (1 + \rho)\, c_1^\alpha(\rho = 0) - \rho\, c_1^\alpha(\rho = -1)$$

$$c_2^\alpha(\rho < 0) = a_2^\alpha + b_2^\alpha$$

$$c_3^\alpha(\rho < 0) = a_3^\alpha + b_3^\alpha$$

$$c_4^\alpha(\rho < 0) = (1 + \rho)\, c_4^\alpha(\rho = 0) - \rho\, c_4^\alpha(\rho = -1)$$

Figures 7.48–7.51 show the results obtained by summing up the RFVs of Fig. 7.39 under the hypothesis of the following correlation coefficients, respectively: $\rho = 0$, $\rho = 1$, $\rho = -1$, and $\rho = 0.6$.

Difference

Let w_A and w_B be the semi-widths of the internal fuzzy variables of A and B, respectively:

$$w_A = \frac{a_3^{\alpha=1} - a_2^{\alpha=1}}{2}$$

$$w_B = \frac{b_3^{\alpha=1} - b_2^{\alpha=1}}{2}$$

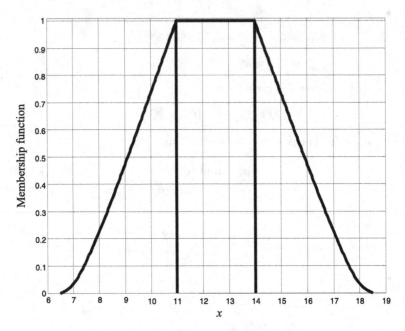

Fig. 7.48. Sum of RFVs A and B of Fig. 7.39 under the hypothesis of total uncorrelation.

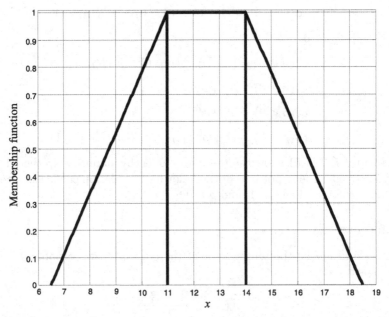

Fig. 7.49. Sum of RFVs A and B of Fig. 7.39 under the hypothesis of total positive correlation.

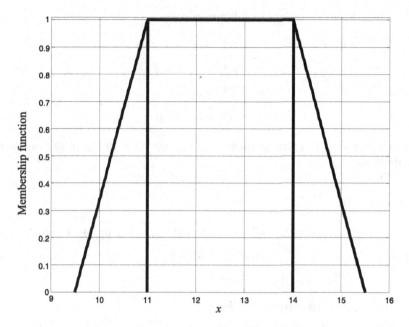

Fig. 7.50. Sum of RFVs A and B of Fig. 7.39 under the hypothesis of total negative correlation.

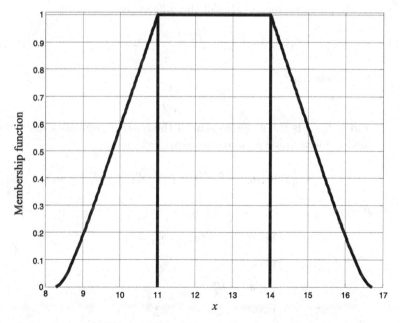

Fig. 7.51. Sum of RFVs A and B of Fig. 7.39 under the hypothesis of $\rho = -0.6$.

Then, let us consider the following:

$$a_{r1}^\alpha = a_1^\alpha + w_A$$
$$a_{r4}^\alpha = a_4^\alpha - w_A$$
$$b_{r1}^\alpha = b_1^\alpha + w_B$$
$$b_{r4}^\alpha = b_4^\alpha - w_B$$

which represent the part of each α-cut associated with the pure random contributions.

Then,

- $\rho = 0$

$$c_1^\alpha(\rho = 0) = c_2^\alpha - \mu_c + k \cdot \text{ext}(a_{r1}^\alpha - b_{r4}^\alpha, g_1^\alpha) + (1 - k)$$
$$\cdot \text{int}(a_{r1}^\alpha - b_{r4}^\alpha, g_1^\alpha)$$
$$c_2^\alpha(\rho = 0) = a_2^\alpha - b_3^\alpha$$
$$c_3^\alpha(\rho = 0) = a_3^\alpha - b_2^\alpha$$
$$c_4^\alpha(\rho = 0) = c_3^\alpha - \mu_c + k \cdot \text{ext}(a_{r4}^\alpha - b_{r1}^\alpha, g_4^\alpha) + (1 - k)$$
$$\cdot \text{int}(a_{r4}^\alpha - b_{r1}^\alpha, g_4^\alpha)$$

where μ_c is the mean value of the difference:

$$\mu_c = \frac{c_2^{\alpha=1} + c_3^{\alpha=1}}{2}$$

k is a constant:

$$k = \frac{1}{\sqrt{2}}$$

and $[g_1^\alpha, g_4^\alpha]$ is the generic α-cut of the normal possibility distribution, having mean value μ_c and standard deviation:

$$\sigma = \frac{1}{3} \min\{\mu_c - a_{r1}^{\alpha=0} + b_{r4}^{\alpha=0}; \ a_{r4}^{\alpha=0} - b_{r1}^{\alpha=0} - \mu_c\},$$

- $\rho = +1$

$$c_1^\alpha(\rho = 1) = c_2^\alpha + s_1 \cdot (a_{r1}^\alpha - b_{r4}^\alpha - \mu_c)$$
$$c_2^\alpha(\rho = 1) = a_2^\alpha - b_3^\alpha$$
$$c_3^\alpha(\rho = 1) = a_3^\alpha - b_2^\alpha$$
$$c_4^\alpha(\rho = 1) = c_3^\alpha + s_4 \cdot (a_{r4}^\alpha - b_{r1}^\alpha - \mu_c)$$

where μ_c is the mean value of the difference,

$$s_1 = \frac{\mu_c - \min\left\{a_{r1}^{\alpha=0} - b_{r1}^{\alpha=0}; a_{r4}^{\alpha=0} - b_{r4}^{\alpha=0}\right\}}{\mu_c - (a_{r1}^{\alpha=0} - b_{r4}^{\alpha=0})}$$

and

$$s_4 = \frac{\max\left\{a_{r1}^{\alpha=0} - b_{r1}^{\alpha=0}; a_{r4}^{\alpha=0} - b_{r4}^{\alpha=0}\right\} - \mu_c}{(a_{r4}^{\alpha=0} - b_{r1}^{\alpha=0}) - \mu_c}$$

- $\rho = -1$

$$c_1^\alpha(\rho = -1) = a_1^\alpha - b_4^\alpha$$
$$c_2^\alpha(\rho = -1) = a_2^\alpha - b_3^\alpha$$
$$c_3^\alpha(\rho = -1) = a_3^\alpha - b_2^\alpha$$
$$c_4^\alpha(\rho = -1) = a_4^\alpha - b_1^\alpha$$

- $0 < \rho < 1$

$$c_1^\alpha(\rho > 0) = (1 - \rho)\, c_1^\alpha(\rho = 0) + \rho\, c_1^\alpha(\rho = 1)$$
$$c_2^\alpha(\rho > 0) = a_2^\alpha - b_3^\alpha$$
$$c_3^\alpha(\rho > 0) = a_3^\alpha - b_2^\alpha$$
$$c_4^\alpha(\rho > 0) = (1 - \rho)\, c_4^\alpha(\rho = 0) + \rho\, c_4^\alpha(\rho = 1)$$

- $-1 < \rho < 0$

$$c_1^\alpha(\rho < 0) = (1 + \rho)\, c_1^\alpha(\rho = 0) - \rho\, c_1^\alpha(\rho = -1)$$
$$c_2^\alpha(\rho < 0) = a_2^\alpha - b_3^\alpha$$
$$c_3^\alpha(\rho < 0) = a_3^\alpha - b_2^\alpha$$
$$c_4^\alpha(\rho < 0) = (1 + \rho)\, c_4^\alpha(\rho = 0) - \rho\, c_4^\alpha(\rho = -1)$$

Figures 7.52–7.54 show the results obtained by subtracting the RFVs of Fig. 7.39 under the hypothesis of the following correlation coefficients, respectively: $\rho = 0$, $\rho = 1$, $\rho = -1$.

Product

Let w_A and w_B be the semi-widths of the internal fuzzy variables of A and B, respectively:

$$w_A = \frac{a_3^{\alpha=1} - a_2^{\alpha=1}}{2}$$

$$w_B = \frac{b_3^{\alpha=1} - b_2^{\alpha=1}}{2}$$

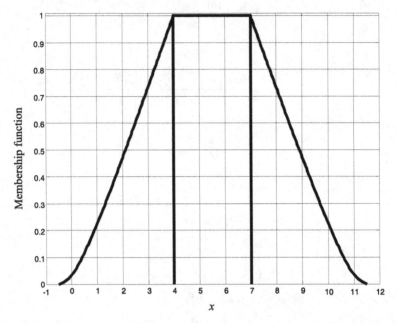

Fig. 7.52. Difference of RFVs A and B of Fig. 7.39 under the hypothesis of no correlation.

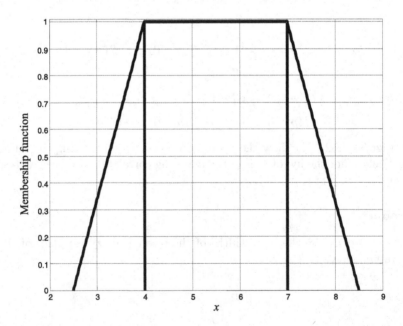

Fig. 7.53. Difference of RFVs A and B of Fig. 7.39 under the hypothesis of total positive correlation.

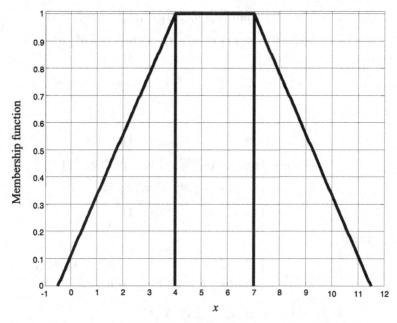

Fig. 7.54. Difference of RFVs RFVs A and B of Fig. 7.39 under the hypothesis of total negative correlation.

Then, let us consider the following:

$$a_{r1}^\alpha = a_1^\alpha + w_A$$

$$a_{r4}^\alpha = a_4^\alpha - w_A$$

$$b_{r1}^\alpha = b_1^\alpha + w_B$$

$$b_{r4}^\alpha = b_4^\alpha - w_B$$

which represent the part of each α-cut associated with the pure random contributions.

Then,

- $\rho = 0$

$$c_1^\alpha(\rho = 0) = c_2^\alpha - \mu_r + c_{r1}^\alpha(\rho = 0)$$

$$c_2^\alpha(\rho = 0) = \min\{a_2^\alpha\ b_2^\alpha; a_2^\alpha\ b_3^\alpha; a_3^\alpha\ b_2^\alpha; a_3^\alpha\ b_3^\alpha\}$$

$$c_3^\alpha(\rho = 0) = \max\{a_2^\alpha\ b_2^\alpha; a_2^\alpha\ b_3^\alpha; a_3^\alpha\ b_2^\alpha; a_3^\alpha\ b_3^\alpha\}$$

$$c_4^\alpha(\rho = 0) = c_3^\alpha - \mu_r + c_{r4}^\alpha(\rho = 0)$$

where

$$c_{r1}^\alpha(\rho = 0) = \min\{a_{r1}^\alpha\ b_{r1}^\alpha; a_{r1}^\alpha\ b_{r4}^\alpha; a_{r4}^\alpha\ b_{r1}^\alpha; a_{r4}^\alpha\ b_{r4}^\alpha\}$$

$$c_{r4}^\alpha(\rho = 0) = \max\{a_{r1}^\alpha\ b_{r1}^\alpha; a_{r1}^\alpha\ b_{r4}^\alpha; a_{r4}^\alpha\ b_{r1}^\alpha; a_{r4}^\alpha\ b_{r4}^\alpha\}$$

and

$$\mu_r = \frac{c_{r1}^{\alpha=1} + c_{r4}^{\alpha=1}}{2} = c_{r1}^{\alpha=1}$$

- $\rho = +1$

$$c_1^\alpha(\rho = 1) = c_2^\alpha - \mu_r + c_{r1}^\alpha(\rho = 1)$$

$$c_2^\alpha(\rho = 1) = \min\{a_2^\alpha\ b_2^\alpha; a_2^\alpha\ b_3^\alpha; a_3^\alpha\ b_2^\alpha; a_3^\alpha\ b_3^\alpha\}$$

$$c_3^\alpha(\rho = 1) = \max\{a_2^\alpha\ b_2^\alpha; a_2^\alpha\ b_3^\alpha; a_3^\alpha\ b_2^\alpha; a_3^\alpha\ b_3^\alpha\}$$

$$c_4^\alpha(\rho = 1) = c_3^\alpha - \mu_r + c_{r4}^\alpha(\rho = 1)$$

It is:

$$\tilde{c}_{r1}^\alpha(\rho = 1) = \begin{cases} x_m^\alpha\ y_m^\alpha & \text{if } a_{r1}^\alpha < x_m^\alpha < a_{r4}^\alpha \\ \min\{a_{r1}^\alpha\ b_{r1}^\alpha; a_{r4}^\alpha\ b_{r4}^\alpha\} & \text{otherwise} \end{cases}$$

$$\tilde{c}_{r4}^\alpha(\rho = 1) = \max\{a_{r1}^\alpha\ b_{r1}^\alpha; a_{r4}^\alpha\ b_{r4}^\alpha\}$$

where, according to Eqs. (7.7) and (7.8),

$$x_m^\alpha = \frac{a_{r1}^\alpha + a_{r4}^\alpha}{4} - \frac{r^\alpha}{4}(b_{r1}^\alpha + b_{r4}^\alpha)$$

$$y_m^\alpha = \frac{b_{r1}^\alpha + b_{r4}^\alpha}{2} + \frac{1}{r^\alpha}\left[x_m^\alpha - \left(\frac{a_{r1}^\alpha + a_{r4}^\alpha}{2}\right)\right] = \frac{b_{r1}^\alpha + b_{r4}^\alpha}{4} - \frac{a_{r1}^\alpha + a_{r4}^\alpha}{4\ r^\alpha}$$

$$r^\alpha = \frac{a_{r4}^\alpha - a_{r1}^\alpha}{b_{r4}^\alpha - b_{r1}^\alpha}$$

$$\mu_r = \frac{c_{r1}^{\alpha=1} + c_{r4}^{\alpha=1}}{2} = c_{r1}^{\alpha=1}$$

If the above values of $\tilde{c}_{r1}^\alpha$ and $\tilde{c}_{r4}^\alpha$ define a membership function, it is $c_{r1}^\alpha = \tilde{c}_{r1}^\alpha$ and $c_{r4}^\alpha = \tilde{c}_{r4}^\alpha$. Otherwise, a correction must be applied, that can be algorithmically defined, for each couple of α-cuts at levels α and $\alpha + d\alpha$, $\forall \alpha$, as:

$$\textit{if } \tilde{c}_{r1}^{\alpha+d\alpha} < \tilde{c}_{r1}^\alpha$$

$$\textit{then } c_{r1}^{\alpha+d\alpha} = \tilde{c}_{r1}^\alpha$$

$$\textit{else } c_{r1}^{\alpha+d\alpha} = \tilde{c}_{r1}^{\alpha+d\alpha}$$

$$\textit{if } \tilde{c}_{r4}^{\alpha+d\alpha} > \tilde{c}_{r4}^\alpha$$

$$\textit{then } c_{r4}^{\alpha+d\alpha} = \tilde{c}_{r4}^\alpha$$

$$\textit{else } c_{r4}^{\alpha+d\alpha} = \tilde{c}_{r4}^{\alpha+d\alpha}$$

- $\rho = -1$

$$c_1^\alpha(\rho = -1) = c_2^\alpha - \mu_r + c_{r1}^\alpha(\rho = -1)$$

$$c_2^\alpha(\rho = -1) = \min\{a_2^\alpha\ b_2^\alpha; a_2^\alpha\ b_3^\alpha; a_3^\alpha\ b_2^\alpha; a_3^\alpha\ b_3^\alpha\}$$

$$c_3^\alpha(\rho = -1) = \max\{a_2^\alpha\ b_2^\alpha; a_2^\alpha\ b_3^\alpha; a_3^\alpha\ b_2^\alpha; a_3^\alpha\ b_3^\alpha\}$$

$$c_4^\alpha(\rho = -1) = c_3^\alpha - \mu_r + c_{r4}^\alpha(\rho = -1)$$

It is:

$$\tilde{c}_{r1}^\alpha(\rho = -1) = \min\{a_{r1}^\alpha\ b_{r4}^\alpha, a_{r4}^\alpha\ b_{r1}^\alpha\}$$

$$\tilde{c}_{r4}^\alpha(\rho = -1) = \begin{cases} x_M^\alpha\ y_M^\alpha & \text{if } a_{r1}^\alpha < x_M^\alpha < a_{r4}^\alpha \\ \max\{a_{r1}^\alpha\ b_{r4}^\alpha, a_{r4}^\alpha\ b_{r1}^\alpha\} & \text{otherwise} \end{cases}$$

where, according to Eqs. (7.12) and (7.13),

$$x_M^\alpha = \frac{a_{r1}^\alpha + a_{r4}^\alpha}{4} + \frac{r^\alpha}{4}\left(b_{r1}^\alpha + b_{r4}^\alpha\right)$$

$$y_M^\alpha = \frac{b_{r1}^\alpha + b_{r4}^\alpha}{2} - \frac{1}{r^\alpha}\left[x_M^\alpha - \left(\frac{a_{r1}^\alpha + a_{r4}^\alpha}{2}\right)\right] = \frac{b_{r1}^\alpha + b_{r4}^\alpha}{4} + \frac{a_{r1}^\alpha + a_{r4}^\alpha}{4\ r^\alpha}$$

$$r^\alpha = \frac{a_{r4}^\alpha - a_{r1}^\alpha}{b_{r4}^\alpha - b_{r1}^\alpha}$$

$$\mu_r = \frac{c_{r1}^{\alpha=1} + c_{r4}^{\alpha=1}}{2} = c_{r1}^{\alpha=1}$$

If the above values of $\tilde{c}_{r1}^\alpha$ and $\tilde{c}_{r4}^\alpha$ define a membership function, it is $c_{r1}^\alpha = \tilde{c}_{r1}^\alpha$ and $c_{r4}^\alpha = \tilde{c}_{r4}^\alpha$. Otherwise, a correction must be applied, that can be algorithmically defined, for each couple of α-cuts at levels α and $\alpha + d\alpha, \forall\alpha$, as:

$$\textit{if } \tilde{c}_{r1}^{\alpha+d\alpha} < \tilde{c}_{r1}^\alpha$$
$$\textit{then } c_{r1}^{\alpha+d\alpha} = \tilde{c}_{r1}^\alpha$$
$$\textit{else } c_{r1}^{\alpha+d\alpha} = \tilde{c}_{r1}^{\alpha+d\alpha}$$
$$\textit{if } \tilde{c}_{r4}^{\alpha+d\alpha} > \tilde{c}_{r4}^\alpha$$
$$\textit{then } c_{r4}^{\alpha+d\alpha} = \tilde{c}_{r4}^\alpha$$
$$\textit{else } c_{r4}^{\alpha+d\alpha} = \tilde{c}_{r4}^{\alpha+d\alpha}$$

- $0 < \rho < 1$

$$c_1^\alpha(\rho > 0) = (1 - \rho)\ c_1^\alpha(\rho = 0) + \rho\ c_1^\alpha(\rho = 1)$$

$$c_2^\alpha(\rho > 0) = \min\{a_2^\alpha\ b_2^\alpha; a_2^\alpha\ b_3^\alpha; a_3^\alpha\ b_2^\alpha; a_3^\alpha\ b_3^\alpha\}$$

$$c_3^\alpha(\rho > 0) = \max\{a_2^\alpha\ b_2^\alpha; a_2^\alpha\ b_3^\alpha; a_3^\alpha\ b_2^\alpha; a_3^\alpha\ b_3^\alpha\}$$

$$c_4^\alpha(\rho > 0) = (1 - \rho)\ c_4^\alpha(\rho = 0) + \rho\ c_4^\alpha(\rho = 1)$$

- $-1 < \rho < 0$

$$c_1^\alpha(\rho < 0) = (1 + \rho)\, c_1^\alpha(\rho = 0) - \rho\, c_1^\alpha(\rho = -1)$$
$$c_2^\alpha(\rho < 0) = \min\{a_2^\alpha\, b_2^\alpha;\, a_2^\alpha\, b_3^\alpha;\, a_3^\alpha\, b_2^\alpha;\, a_3^\alpha\, b_3^\alpha\}$$
$$c_3^\alpha(\rho < 0) = \max\{a_2^\alpha\, b_2^\alpha;\, a_2^\alpha\, b_3^\alpha;\, a_3^\alpha\, b_2^\alpha;\, a_3^\alpha\, b_3^\alpha\}$$
$$c_4^\alpha(\rho < 0) = (1 + \rho)\, c_4^\alpha(\rho = 0) - \rho\, c_4^\alpha(\rho = -1)$$

Figures 7.55 and 7.56 show the results obtained by multiplying the RFVs of Fig. 7.39 under the hypothesis of the following correlation coefficients, respectively: $\rho = 0$, $\rho = 1$, and $\rho = -1$. It can be noted that, for this particular example, the results obtained for $\rho = 0$ and $\rho = 1$ are the same.

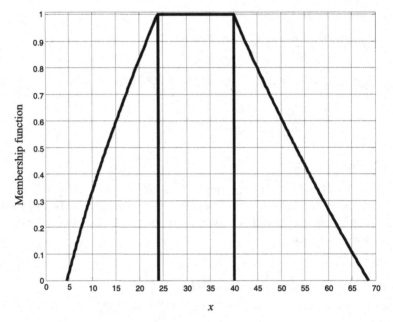

Fig. 7.55. Product of RFVs A and B of Fig. 7.39 under the hypothesis of total uncorrelation. The same result is also obtained under the hypothesis of total positive correlation.

Division

Let w_A and w_B be the semi-widths of the internal fuzzy variables of A and B, respectively:

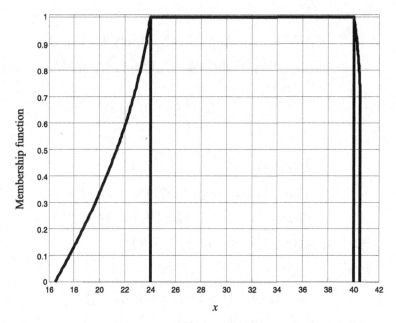

Fig. 7.56. Product of RFVs A and B of Fig. 7.39 under the hypothesis of total negative correlation.

$$w_A = \frac{a_3^{\alpha=1} - a_2^{\alpha=1}}{2}$$

$$w_B = \frac{b_3^{\alpha=1} - b_2^{\alpha=1}}{2}$$

Then, let us consider the following:

$$a_{r1}^{\alpha} = a_1^{\alpha} + w_A$$

$$a_{r4}^{\alpha} = a_4^{\alpha} - w_A$$

$$b_{r1}^{\alpha} = b_1^{\alpha} + w_B$$

$$b_{r4}^{\alpha} = b_4^{\alpha} - w_B$$

which represent the part of each α-cut associated with the pure random contributions.

- $\rho = 0$

$$c_1^{\alpha}(\rho = 0) = c_2^{\alpha} - \mu_r + c_{r1}^{\alpha}(\rho = 0)$$

$$c_2^{\alpha}(\rho = 0) = \min\{a_2^{\alpha}/b_2^{\alpha}; a_2^{\alpha}/b_3^{\alpha}; a_3^{\alpha}/b_2^{\alpha}; a_3^{\alpha}/b_3^{\alpha}\}$$

$$c_3^{\alpha}(\rho = 0) = \max\{a_2^{\alpha}/b_2^{\alpha}; a_2^{\alpha}/b_3^{\alpha}; a_3^{\alpha}/b_2^{\alpha}; a_3^{\alpha}/b_3^{\alpha}\}$$

$$c_4^{\alpha}(\rho = 0) = c_3^{\alpha} - \mu_r + c_{r4}^{\alpha}(\rho = 0)$$

where

$$c_{r1}^\alpha(\rho = 0) = \min\{a_{r1}^\alpha/b_{r1}^\alpha; a_{r1}^\alpha/b_{r4}^\alpha; a_{r4}^\alpha/b_{r1}^\alpha; a_{r4}^\alpha/b_{r4}^\alpha\}$$

$$c_{r4}^\alpha(\rho = 0) = \max\{a_{r1}^\alpha\, b_{r1}^\alpha; a_{r1}^\alpha\, b_{r4}^\alpha; a_{r4}^\alpha\, b_{r1}^\alpha; a_{r4}^\alpha\, b_{r4}^\alpha\}$$

and

$$\mu_r = \frac{c_{r1}^{\alpha=1} + c_{r4}^{\alpha=1}}{2} = c_{r1}^{\alpha=1}$$

- $\rho = +1$

$$c_1^\alpha(\rho = 1) = c_2^\alpha - \mu_r + c_{r1}^\alpha(\rho = 1)$$

$$c_2^\alpha(\rho = 1) = \min\{a_2^\alpha/b_2^\alpha; a_2^\alpha/b_3^\alpha; a_3^\alpha/b_2^\alpha; a_3^\alpha/b_3^\alpha\}$$

$$c_3^\alpha(\rho = 1) = \max\{a_2^\alpha/b_2^\alpha; a_2^\alpha/b_3^\alpha; a_3^\alpha/b_2^\alpha; a_3^\alpha/b_3^\alpha\}$$

$$c_4^\alpha(\rho = 1) = c_3^\alpha - \mu_r + c_{r4}^\alpha(\rho = 1)$$

It is:

$$\tilde{c}_{r1}^\alpha(\rho = 1) = \min\{a_{r1}^\alpha/b_{r1}^\alpha; a_{r4}^\alpha/b_{r4}^\alpha\}$$

$$\tilde{c}_{r4}^\alpha(\rho = 1) = \max\{a_{r1}^\alpha/b_{r1}^\alpha; a_{r4}^\alpha/b_{r4}^\alpha\}$$

and

$$\mu_r = \frac{c_{r1}^{\alpha=1} + c_{r4}^{\alpha=1}}{2} = c_{r1}^{\alpha=1}$$

If the above values of $\tilde{c}_{r1}^\alpha$ and $\tilde{c}_{r4}^\alpha$ define a membership function, it is $c_{r1}^\alpha = \tilde{c}_{r1}^\alpha$ and $c_{r4}^\alpha = \tilde{c}_{r4}^\alpha$. Otherwise, a correction must be applied, that can be algorithmically defined, for each couple of α-cuts at levels α and $\alpha + d\alpha, \forall \alpha$, as:

$$\textit{if} \quad \tilde{c}_{r1}^{\alpha+d\alpha} < \tilde{c}_{r1}^\alpha$$

$$\textit{then} \quad c_{r1}^{\alpha+d\alpha} = \tilde{c}_{r1}^\alpha$$

$$\textit{else} \quad c_{r1}^{\alpha+d\alpha} = \tilde{c}_{r1}^{\alpha+d\alpha}$$

$$\textit{if} \quad \tilde{c}_{r4}^{\alpha+d\alpha} > \tilde{c}_{r4}^\alpha$$

$$\textit{then} \quad c_{r4}^{\alpha+d\alpha} = \tilde{c}_{r4}^\alpha$$

$$\textit{else} \quad c_{r4}^{\alpha+d\alpha} = \tilde{c}_{r4}^{\alpha+d\alpha}$$

- $\rho = -1$

$$c_1^\alpha(\rho = -1) = c_2^\alpha - \mu_r + c_{r1}^\alpha(\rho = -1)$$

$$c_2^\alpha(\rho = -1) = \min\{a_2^\alpha/b_2^\alpha; a_2^\alpha/b_3^\alpha; a_3^\alpha/b_2^\alpha; a_3^\alpha/b_3^\alpha\}$$

$$c_3^\alpha(\rho = -1) = \max\{a_2^\alpha/b_2^\alpha; a_2^\alpha/b_3^\alpha; a_3^\alpha/b_2^\alpha; a_3^\alpha/b_3^\alpha\}$$

$$c_4^\alpha(\rho = -1) = c_3^\alpha - \mu_r + c_{r4}^\alpha(\rho = -1)$$

It is:

$$\tilde{c}_{r1}^{\alpha}(\rho = -1) = \min\{a_{r1}^{\alpha}/b_{r4}^{\alpha}; a_{r4}^{\alpha}/b_{r1}^{\alpha}\}$$

$$\tilde{c}_{r4}^{\alpha}(\rho = -1) = \max\{a_{r1}^{\alpha}/b_{r4}^{\alpha}; a_{r4}^{\alpha}/b_{r1}^{\alpha}\}$$

and

$$\mu_r = \frac{c_{r1}^{\alpha=1} + c_{r4}^{\alpha=1}}{2} = c_{r1}^{\alpha=1}$$

If the above values of $\tilde{c}_{r1}^{\alpha}$ and $\tilde{c}_{r4}^{\alpha}$ define a membership function, it is $c_{r1}^{\alpha} = \tilde{c}_{r1}^{\alpha}$ and $c_{r4}^{\alpha} = \tilde{c}_{r4}^{\alpha}$. Otherwise, a correction must be applied, that can be algorithmically defined, for each couple of α-cuts at levels α and $\alpha + d\alpha, \forall \alpha$, as:

$$if \quad \tilde{c}_{r1}^{\alpha+d\alpha} < \tilde{c}_{r1}^{\alpha}$$
$$then \quad c_{r1}^{\alpha+d\alpha} = \tilde{c}_{r1}^{\alpha}$$
$$else \quad c_{r1}^{\alpha+d\alpha} = \tilde{c}_{r1}^{\alpha+d\alpha}$$
$$if \quad \tilde{c}_{r4}^{\alpha+d\alpha} > \tilde{c}_{r4}^{\alpha}$$
$$then \quad c_{r4}^{\alpha+d\alpha} = \tilde{c}_{r4}^{\alpha}$$
$$else \quad c_{r4}^{\alpha+d\alpha} = \tilde{c}_{r4}^{\alpha+d\alpha}$$

- $0 < \rho < 1$

$$c_1^{\alpha}(\rho > 0) = (1 - \rho)\, c_1^{\alpha}(\rho = 0) + \rho\, c_1^{\alpha}(\rho = 1)$$
$$c_2^{\alpha}(\rho > 0) = \min\{a_2^{\alpha}/b_2^{\alpha}; a_2^{\alpha}/b_3^{\alpha}; a_3^{\alpha}/b_2^{\alpha}; a_3^{\alpha}/b_3^{\alpha}\}$$
$$c_3^{\alpha}(\rho > 0) = \max\{a_2^{\alpha}/b_2^{\alpha}; a_2^{\alpha}/b_3^{\alpha}; a_3^{\alpha}/b_2^{\alpha}; a_3^{\alpha}/b_3^{\alpha}\}$$
$$c_4^{\alpha}(\rho > 0) = (1 - \rho)\, c_4^{\alpha}(\rho = 0) + \rho\, c_4^{\alpha}(\rho = 1)$$

- $-1 < \rho < 0$

$$c_1^{\alpha}(\rho < 0) = (1 + \rho)\, c_1^{\alpha}(\rho = 0) - \rho\, c_1^{\alpha}(\rho = -1)$$
$$c_2^{\alpha}(\rho < 0) = \min\{a_2^{\alpha}/b_2^{\alpha}; a_2^{\alpha}/b_3^{\alpha}; a_3^{\alpha}/b_2^{\alpha}; a_3^{\alpha}/b_3^{\alpha}\}$$
$$c_3^{\alpha}(\rho < 0) = \max\{a_2^{\alpha}/b_2^{\alpha}; a_2^{\alpha}/b_3^{\alpha}; a_3^{\alpha}/b_2^{\alpha}; a_3^{\alpha}/b_3^{\alpha}\}$$
$$c_4^{\alpha}(\rho < 0) = (1 + \rho)\, c_4^{\alpha}(\rho = 0) - \rho\, c_4^{\alpha}(\rho = -1)$$

Figures 7.57 and 7.58 show the results obtained by dividing the RFVs of Fig. 7.39 under the hypothesis of the following correlation coefficients, respectively: $\rho = 0$, $\rho = 1$, and $\rho = -1$. It can be noted that, for this particular example, the results obtained for $\rho = 0$ and $\rho = -1$ are the same.

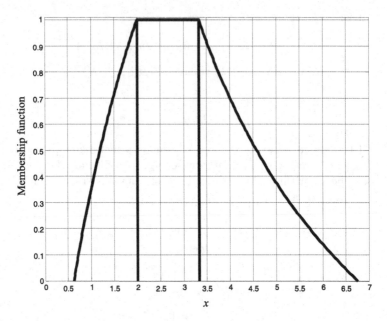

Fig. 7.57. Division of RFVs A and B of Fig. 7.39 under the hypothesis of total uncorrelation. The same result is also obtained under the hypothesis of total negative correlation.

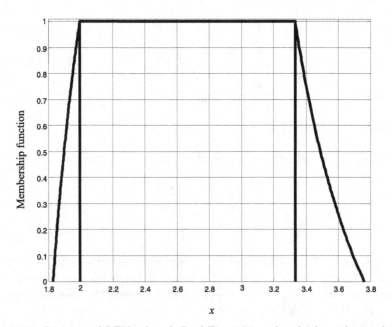

Fig. 7.58. Division of RFVs A and B of Fig. 7.39 under the hypothesis of total positive correlation.

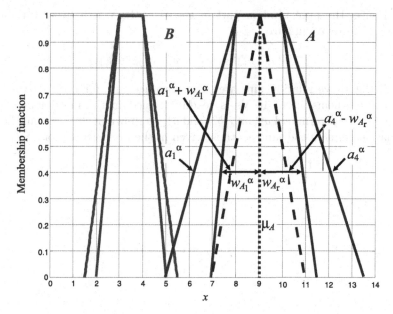

Fig. 7.59. Determination of the pure random part in the more generic case.

7.3.3 General situation

The more general situation, where the considered RFVs show both a random and a non-random part and at least one of the considered RFVs has an internal membership function that is not rectangular, as shown in Fig. 7.40, requires a generalization of the combination rules given in the previous section.[15]

In this case, it must be considered that the width of the internal intervals, intercepted, on each α-cut, by the internal membership function, is no longer constant but varies with level α. Moreover, the internal fuzzy variable could be asymmetric, so that, with respect to the mean value, the width of the left side of the interval is different from the width of the right side.

In particular, if an RFV A is considered, as shown in Fig. 7.59, the width of the left part is

$$w_{Al}^{\alpha} = \mu_A - a_2^{\alpha}$$

and the width of the right part is

$$w_{Ar}^{\alpha} = a_3^{\alpha} - \mu_A$$

---

[15] Let us remember that the simpler situation has been first considered due to the fact that most measurement results are represented by RFVs like the ones in Fig. 7.39, whereas RFVs like the ones in Fig. 7.40 are seldom met in practice.

Hence, the pure random part of the given RFV is a fuzzy variable (of type 1) whose α-cuts are

$$[a^\alpha_{r1}, a^\alpha_{r4}] = [a^\alpha_1 + w^\alpha_{Al}; a^\alpha_4 - w^\alpha_{Ar}]$$

If the above relationships are taken into account, for both RFVs A and B, the mathematics of RFVs is readily defined in the more general case. Infact, once determined $a^\alpha_{r1}, a^\alpha_{r4}, b^\alpha_{r1}$ and b^α_{r4} as shown in this section, the final equations remain the same as those given in the previous section.

8

Representation of Random–Fuzzy Variables

To allow an efficient handling of the random–fuzzy variables also from the numerical point of view, it is necessary to establish an efficient method for their representation. The method should be chosen so as to allow an easy storage and a reduced memory usage, and an easy implementation of the mathematical operations and functions defined for the random–fuzzy variables.

As shown in the previous chapters, all operations and functions for the random–fuzzy variables have been defined referring to the concept of α-cut and must be applied to each α-cut of the random–fuzzy variables themselves. Therefore, the more intuitive and natural way to numerically represent a random–fuzzy variable is to consider its α-cuts.

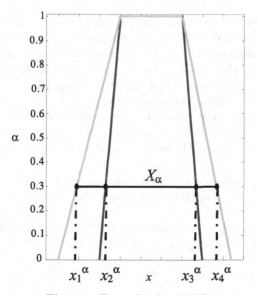

Fig. 8.1. Example of an RFV.

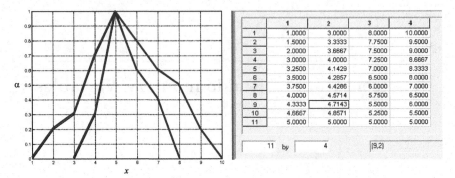

Fig. 8.2. Representation of a random–fuzzy variable in terms of a $[11 \times 4]$ matrix. Each row corresponds to an α-cut. The α-cuts are taken at values $\alpha = 0, 0.1, 0.2, \ldots, 0.9, 1$.

Let us remember that, as shown in Fig. 8.1, each α-cut of an RFV is described by four numbers:

$$X_\alpha = \{x_1^\alpha, x_2^\alpha, x_3^\alpha, x_4^\alpha\}$$

This equation suggests to represent an RFV by a matrix, containing the above four numbers of each different α-cut X_α in its rows.

Hence, by definition, this matrix has four columns, each one containing one of the four values defined by the α-cut: The 1st column contains value x_1^α, the 2nd column contains value x_2^α, the 3rd column contains value x_3^α, and the 4th column contains value x_4^α.

On the other hand, the number of rows is not a priori given and can be chosen by the operator. Of course, the greater is the number of rows, the higher is the accuracy with which the random–fuzzy variable is represented. Obviously, both the required memory and the computational burden grow together with the number of rows, that is, the number of α-cuts with which the random–fuzzy variable is represented.

A good compromise, which ensures both a very good resolution and a low computational burden, in most practical applications, is the use of 101 rows, that is 101 α-cuts. Under this assumption, each random–fuzzy variable is represented by a 101 × 4 matrix.

If this representation is followed, operations between random–fuzzy variables become simply operations between matrices, so that their processing becomes easy.

Figure 8.2 shows an example of a random–fuzzy variable and its matrix representation in the simplified case, for the sake of brevity, of only 11 rows.

9

Decision-Making Rules with Random–Fuzzy Variables

Practically all measurement processes end up in making a decision, according to the result of a comparison between the measurement result and another given value, either a threshold or another measurement result. In the simplest situation, when the measurement result has to be compared with a threshold and the measurement uncertainty is taken into account by representing the measurement result with a random–fuzzy variable (RFV), this comparison is performed between an RFV and a scalar value. In more complicated situations, when the measurement result has to be compared with another measurement value, the comparison is performed between two RFVs.

Examples of these cases are reported in Figs. 9.1 and 9.2. In these figures, the RFV is represented with only its external membership function for the sake of clearness, as if it were a type 1 fuzzy variable. It is worth noting, however, that this case can be considered as the more general one. In fact, when a measurement result has to be compared with a threshold, or with another measurement result, the effects of all uncertainty contributions must be considered in this comparison. This means that, if the measurement result is represented by an RFV, only the external membership function, which takes into account the whole effects of the different uncertainty contributions as a whole, has to be considered. Hence, in this respect, it can be considered as a fuzzy variable of type 1.

Figure 9.1 shows the three different situations that may occur when comparing a fuzzy variable with a scalar value: (a) The RFV is greater than the threshold, (b) the RFV is lower than the threshold, and (c) the RFV crosses the threshold. It can be readily understood that cases a and b do not show any particular problem. In fact, the threshold is lower (or greater) than all possible values that can be assumed by the measurement result. On the other hand, case c is more complicated, because the threshold falls within the interval of the possible values that can be assumed by the measurement result.

Figure 9.2 shows the three different situations that may occur when comparing two RFVs. Similar considerations can be done in this case. It can be

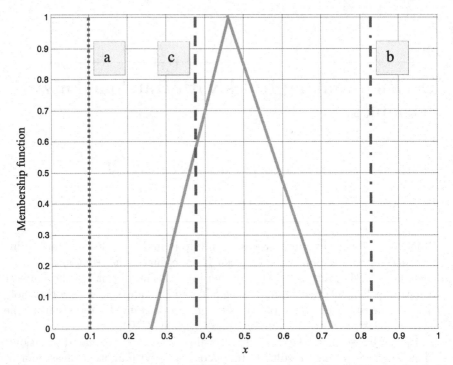

Fig. 9.1. Comparison between an RFV and a scalar value. Three different threshold positions show the possible situations that may occur: (a) The RFV is above the threshold, (b) the RFV is below the threshold, and (c) the RFV crosses the threshold.

easily understood, however, that case c in Fig. 9.2 is surely more complex to handle than the corresponding one in Fig. 9.1.

Figure 9.2 shows also a more general case: the comparison of a measurement result with another measurement result. In this case, however, the measurement uncertainty that affects the second measurement result is much smaller than the uncertainty affecting the first one.

A more general and complex situation occurs when the two measurement results that have to be compared are affected by similar uncertainties, as shown in Fig. 9.3.

The above discussion leads one to conclude that it is necessary to define a suitable ordering rule, for RFVs, in order to attain a unique, unambiguous result from the comparison. In fact, in many practical situations, it is not possible to doubtfully abstain from taking a decision, and a decision must be somehow taken even in the worst cases, like the one, for instance, in Fig. 9.3c.

Moreover, it can be also easily recognized that different grades of belief can be assigned to the statement: 'RFV A is greater (lower) than RFV B', according, for instance, to the three different situations depicted in Fig. 9.3.

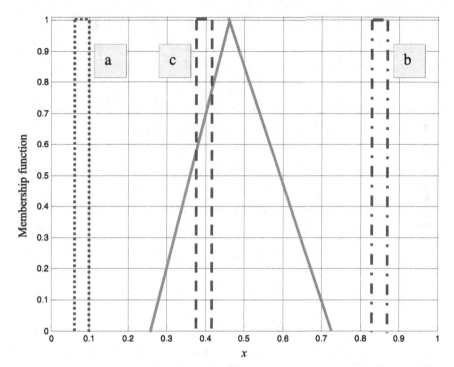

Fig. 9.2. Comparison between two RFVs. The RFV in the solid line represents the measurement result that must be compared with a threshold with uncertainty, whose RFVs in (a) dotted line, (b) dash-dotted line, and (c) dashed line show three different possible positions. Cases a, b, and c show the possible situations that may occur: (a) The RFV is greater than the threshold with uncertainty, (b) the RFV is lower than the threshold with uncertainty, and (c) the RFV partially overlaps the threshold with uncertainty.

This requires that each decision is associated with a given positive number, which provides the grade of belief of the decision itself. This number, called the *credibility factor*, takes values in the range [0,1], where 1 means that there is total evidence in support to the decision, which is hence taken with full certainty, and values lower than 1 mean that only partial evidence supports the decision.

Many different methods are available in the literature to compare fuzzy variables. However, all different methods have been proposed essentially to solve economic problems, and not specifically those in the measurement field. Therefore, in the first part of this chapter, only three methods available in the literature are presented and discussed. These three methods, in fact, are the only ones that seem to be suitable for our aim, whereas all other ones are totally unsuitable; hence, it is not worthwhile investigating them. At last, in the second part of this chapter, a new method is presented.

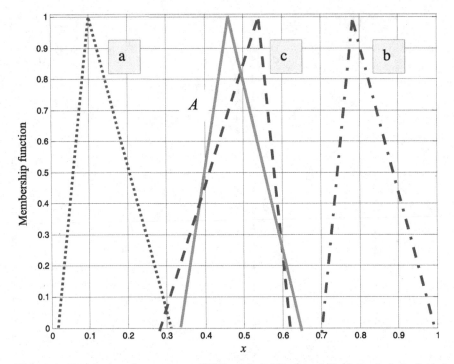

Fig. 9.3. Comparison between two RFVs A (solid line) and B (dotted, dash-dotted and dashed line, respectively). Three different situations may occur: (a) A is greater than B, (b) A is lower than B, and (c) A and B are partially overlapping.

9.1 The available methods

9.1.1 Yager method

The Yager method is based on the Yager area, defined in Chapter 6, for a fuzzy variable A:

$$Y_A = \int_0^1 y_A(\alpha) d\alpha$$

where $y_A(\alpha)$ is the function of the middle points of the various α-cuts of A, at the different levels α.

Given two fuzzy variables A and B, and evaluating the Yager areas Y_A and Y_B, Yager method establishes that $A > B$ if $Y_A > Y_B$ and $A < B$ if $Y_A < Y_B$ [CH92].

An example of this method is reported in Fig. 9.4, where two generic fuzzy variables A and B are considered. In this figure, $y_A(\alpha)$ and $y_B(\alpha)$ are the functions of the middle points of the various α-cuts, at the different levels α, of A and B, respectively, whereas the dashed and gray areas are the Yager areas Y_A and Y_B, respectively.

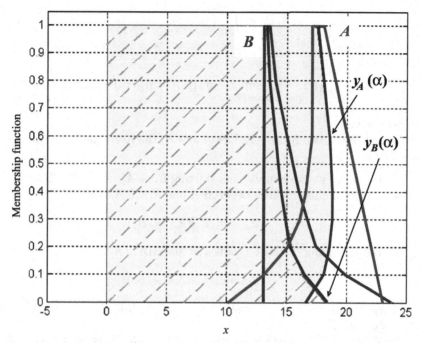

Fig. 9.4. Application of the Yager method on fuzzy variables A and B. Y_A and Y_B are, from a numerical point of view, represented by the gray and dashed areas, respectively.

For this example, it is $Y_A = 18.19$ and $Y_B = 14.59$; hence, the Yager method states that $A > B$.

The Yager method bases its decision on the values of the Yager areas of the two fuzzy variables. This means that, as far as this method is concerned, each fuzzy variable is fully described by a single scalar value. Moreover, this scalar value is evaluated, for each fuzzy variable, independently from the other one. Hence, it can be stated that the comparison is not performed on the two fuzzy variables, but only on one of their characteristics.

Representing a fuzzy variable with a single scalar value does not seem to be fully appropriate. This leads one to conclude that, even if the Yager method provides a correct result, because, in the example of Fig. 9.4, it can be also intuitively stated that $A > B$, a more complex method is required.

Moreover, the Yager method does not associate any credibility factor to the final decision and this is another reason to discard it for our purpose.

9.1.2 Kerre method

The Kerre method is based on the Hamming distance and the fuzzy operator fuzzy-max, as defined in Chapter 6 for two generic fuzzy variables A and B:

$$d(A, B) = \int_{-\infty}^{+\infty} |\mu_A(x) - \mu_B(x)| dx$$

and

$$\mu_{\text{MAX}(A,B)}(z) = \sup_{z=\max\{x,y\}} \min\{\mu_A(x), \mu_B(x)\} \quad \forall x, y, z \in X$$

respectively.

In fact, given two fuzzy variables A and B, the Kerre method requires one to evaluate:

- the fuzzy variable fuzzy-max starting from A and B: MAX(A, B);
- the Hamming distance between A and MAX(A, B):

$$d_A = d(A, \text{MAX}(A, B))$$

- the Hamming distance between B and MAX(A, B):

$$d_B = d(B, \text{MAX}(A, B))$$

Then, the Kerre method establishes that $A > B$ if $d_A < d_B$ and $A < B$ if $d_A > d_B$ [CH92].

An example of this method is reported in Fig. 9.5, where the two considered fuzzy variables are the same as those in Fig. 9.4. The Hamming distance d_A is represented by the dashed area, whereas the Hamming distance d_B is represented by the dotted area. For this example, it is $d_A = 0.164$ and $d_B = 4.5794$; and hence, this method establishes that $A > B$.

By comparing the Kerre method with the Yager method, defined in the previous section, it can be stated that both methods provide the same result. Moreover, the Kerre method does not simply compare two scalar values, each one associated with its own fuzzy variable (as the Yager method does), but it really compares the two fuzzy variables. In fact, it evaluates two values, which depend on the two considered fuzzy variables. Therefore, it can be concluded that the Kerre method is more reliable than the Yager one.

Also in this case, no credibility factor is associated with the final decision.

9.1.3 Nakamura method

The Nakamura method is more complex than both of the previous ones. Similarly to the Kerre method, it is based on the Hamming distance, recalled above, but it also requires the evaluation of the fuzzy variables fuzzy-min, greatest upper set and greatest lower set, as defined in Chapter 6:

$$\mu_{\text{MIN}(A,B)}(z) = \inf_{z=\max\{x,y\}} \min\{\mu_A(x), \mu_B(x)\} \quad \forall x, y, z \in X$$

$$\mu_{A^+}(x) = \max_{y \leq x} \mu_A(y)$$

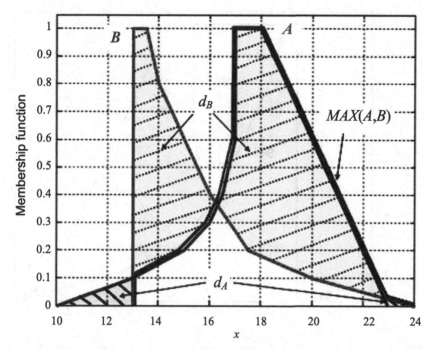

Fig. 9.5. Application of the Kerre method on fuzzy variables A and B. d_A and d_B are, from a numerical point of view, represented by the dashed and dotted areas, respectively. Fuzzy variable MAX(A, B) is also shown.

and

$$\mu_{A^-}(x) = \max_{y \geq x} \mu_A(y)$$

respectively.

Given two fuzzy variables A and B, the Nakamura method requires one to evaluate:

- the greatest upper set and greatest lower set of both variables A and B: A^+, A^-, B^+, and B^-;
- the variable fuzzy-min between A^- and B^- : MIN(A^-, B^-);
- the variable fuzzy-min between A^+ and B^+ : MIN(A^+, B^+);
- the Hamming distance between A^- and MIN(A^-, B^-):

$$d_\tau(A, B) = d(A^-, \text{MIN}(A^-, B^-))$$

- the Hamming distance between A^+ and MIN(A^+, B^+):

$$d_\omega(A, B) = d(A^+, \text{MIN}(A^+, B^+))$$

- the Hamming distance between B^- and MIN(A^-, B^-):

$$d_\xi(A, B) = d(B^-, \text{MIN}(A^-, B^-))$$

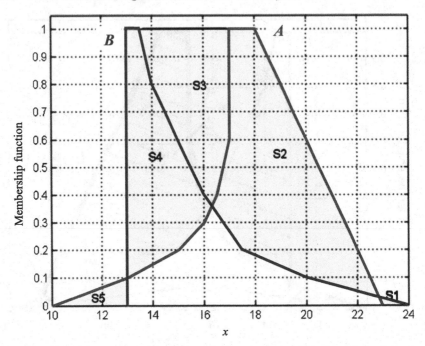

Fig. 9.6. Application of the Nakamura method and identification of the five areas.

- the Hamming distance between B^+ and $\mathrm{MIN}(A^+, B^+)$:

$$d_\psi(A, B) = d(B^+, \mathrm{MIN}(A^+, B^+))$$

- the following two indexes, named preference indexes:

$$P(A, B) = \frac{w\ d_\xi(A, B) + (1 - w)\ d_\psi(A, B)}{w[d_\omega(A, B) + d_\xi(A, B)] + (1 - w)[d_\tau(A, B) + d_\psi(A, B)]}$$

$$P(B, A) = 1 - P(A, B)$$

These two indexes depend on a weight w, whose values can be chosen in a range from 0 and 1.

Then, the Nakamura method establishes that $A > B$ if $P(A, B) > P(B, A)$ and $A < B$ if $P(A, B) < P(B, A)$ [CH92].

Let us now consider the same variables as those in Figs. 9.4 and 9.5. Following the requirements of this method, we obtain that the Hamming distances $d_\tau(A, B), d_\omega(A, B), d_\xi(A, B)$, and $d_\psi(A, B)$ are a suitable combination of the five areas shown in Fig. 9.6 as $S1, S2, S3, S4$, and $S5$.

In particular, it is

$$d_\tau(A, B) = S3 + S4$$
$$d_\omega(A, B) = S2 + S3$$

$$d_\xi(A, B) = S5$$
$$d_\psi(A, B) = S1$$

In the considered example, it is

$$d_\tau(A, B) = 3.062$$
$$d_\omega(A, B) = 4.341$$
$$d_\xi(A, B) = 0.15$$
$$d_\psi(A, B) = 0.014$$

Once these values are known, the last step is the determination of the preference indexes. However, as stated, this passage requires the choice of a weight w. If $w = 0.5$ is considered, it follows that $P(A, B) = 0.026$ and $P(B, A) = 0.974$. Hence, the Nakamura method concludes that $A < B$. The same result is obtained, although with different values for the preference indexes, if any other value is chosen for w.

This result is in contrast with the results obtained by the Yager and the Kerre methods. Hence, this method is not totally reliable, because it can also make wrong decisions.

It could be interesting trying to understand the reasons that lead to a wrong result. Both the Kerre method and the Nakamura method are based on the Hamming distance, and hence, it seems strange that they lead to opposite conclusions.

The reasons why the Nakamura method fails can be searched in the definitions of the different areas, which concur in the evaluation of the preference indexes. In fact, as also shown in Fig. 9.6, area $S3$, which is the area between the two fuzzy variables, is also considered. By comparing Fig. 9.6 with Fig. 9.5, it can be noted that this area is considered by the Nakamura method and not considered by Kerre. Hence, it can be stated that this area is responsible for the wrong result provided by this method. In fact, this area does not belong to any of the two fuzzy variables, and hence, there is no reason why it should be considered in the comparison.

The advantage of the Nakamura method, however, is that it somehow provides a credibility factor, which could be the preference index. The Kerre method and the Yager method, on the other hand, provide the correct result, but they do not provide any credibility factor.

9.2 A specific method

9.2.1 Definition of the credibility coefficients

Although the methods described in the previous sections are somehow capable of assessing whether an RFV is greater or lower than another RFV, they do not succeed in quantifying the grade of belief associated with the selected statement, and hence, they do not succeed in providing an unambiguous rule for ordering RFVs.

On the other side, it is known that the measurement process can be formalized as a morphing mapping from an empirical to a symbolic relational system. This mapping should be capable to preserve the scalar relation of the empirical system into the symbolic one. Therefore, because the quantities that are subject to a measurement process can be put in an ordered relational scale, the same must be done for the measurement results.

It is also known that the result of a measurement represents incomplete knowledge about the measurand, and this incomplete knowledge can be represented by an RFV. Due to this incompleteness, the measurement results cannot be ordered with full certitude, and any adopted rule for their ordering can be effectively used in practice only if it can provide a quantitative estimate of the degree of belief of the obtained ordering. In other words, a credibility factor must be associated with any statement of the kind $A > B$ or $A < B$, being A and B RFVs.

The analysis of the methods briefly recalled in the previous sections led one to conclude that none of them is fully satisfying. However, the advantages and disadvantages of each of them may suggest some new ideas for the definition of a suitable ordering rule.

For instance, our analysis has highlighted that the Hamming distance is able to extract interesting information about the 'distance' between two membership functions. Therefore, let us consider two fuzzy variables A and B and their Hamming distance $d(A, B)$. According to the definition given in Chapter 6, as Fig. 9.7 shows, the Hamming distance is numerically equal to the sum of the four considered areas; that is,

$$d(A, B) = d_1 + d_2 + d_3 + d_4 \qquad (9.1)$$

It is possible to give a specific meaning to each of these areas. In fact, d_1 may represent an indication on how much fuzzy variable A is lower than fuzzy variable B, whereas d_2 may represent an indication on how much fuzzy variable B is lower than fuzzy variable A. On the other hand, d_3 may represent an indication on how much fuzzy variable A is greater than fuzzy variable B, whereas d_4 may represent an indication on how much fuzzy variable B is greater than fuzzy variable A.

Of course, assessing that fuzzy variable A is lower than fuzzy variable B is the same as assessing that fuzzy variable B is greater than fuzzy variable A. Hence, it can be concluded that the sum of areas d_1 and d_4 provides information about the credibility that A is lower than B. In a similar way, it can be stated that the sum of areas d_2 and d_3 provides information about the credibility that A is greater than B.

Figure 9.7 also shows that, if the intersection area between the same fuzzy variables A and B is added to d_1, d_2, d_3, and d_4, the union area is obtained; that is,

$$Un(A, B) = d_1 + d_2 + d_3 + d_4 + Int(A, B) \qquad (9.2)$$

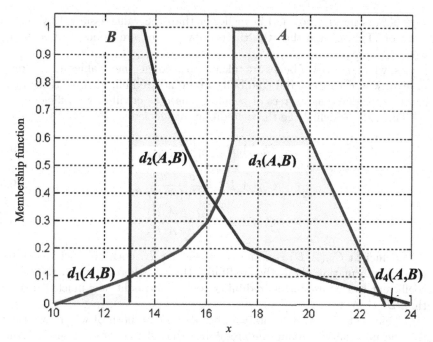

Fig. 9.7. The Hamming distance of the two fuzzy variables A and B is numerically equal to the sum of the four areas d_1, d_2, d_3, and d_4.

Relationship (9.2) can also be rewritten as

$$\frac{d_1 + d_2 + d_3 + d_4 + Int(A,B)}{Un(A,B)} = \frac{d_1 + d_4}{Un(A,B)} + \frac{d_2 + d_3}{Un(A,B)} + \frac{Int(A,B)}{Un(A,B)} = 1 \tag{9.3}$$

Equation (9.3) shows that three ratios can be derived from Eq. (9.2), whose sum is equal to one; in other words, full credibility is assigned to the sum of these three ratios. Moreover, they show some meaningful properties that are worthwhile to investigate further on.

The three ratios $\frac{d_1+d_4}{Un(A,B)}$, $\frac{d_2+d_3}{Un(A,B)}$, and $\frac{Int(A,B)}{Un(A,B)}$ take values in the range [0,1].

In fact, for each ratio, the values contained in both the numerator and the denominator represent specific areas, which, by definition, are positive or, in the limit case in which some area does not exist, null. This proves that the three above ratios cannot assume negative values.

Moreover, for each ratio, the value assumed by the denominator, being the sum of the areas subtended by the standard fuzzy union of the two fuzzy variables, is never zero[1] and never lower than the value assumed by the numerator. This proves that the three above ratios are upper limited to the value 1.

As, when degrees of belief are taken into account, their values always range in $[0,1]$, where 1 means full credibility and 0 no credibility; these three ratios can be considered as degrees of belief and used as credibility coefficients.

Then, let us define the three credibility coefficients:

$$C_{lo}(A, B) = \frac{d_1 + d_4}{Un(A, B)} \tag{9.4}$$

$$C_{gr}(A, B) = \frac{d_2 + d_3}{Un(A, B)} \tag{9.5}$$

$$C_{eq}(A, B) = \frac{Int(A, B)}{Un(A, B)} \tag{9.6}$$

Coefficient $C_{lo}(A, B)$ contains areas d_1 and d_4, both of which, as stated, provide information about the credibility that A is lower than B. Hence, this coefficient can be used as a credibility coefficient about how much A is lower than B.

Coefficient $C_{gr}(A, B)$ contains areas d_2 and d_3, both of which, as stated, provide information about the credibility that A is greater than B. Hence, this coefficient can be used as a credibility coefficient about how much A is greater than B.

Coefficient $C_{eq}(A, B)$ contains the intersection area, which, by definition, provides information about how much the two fuzzy variables overlap.

Let us consider that this ratio assumes a unitary value whenever the two considered fuzzy variables are equal. In fact, two fuzzy variables A and B are equal only if, for every $x \in X$, it is $\mu_A(x) = \mu_B(x)$, that is, if the two membership functions coincide. In this particular situation, the intersection area coincides with the union area, thus leading to the unitary value for coefficient $C_{eq}(A, B)$, which hence represents the credibility coefficient of the statement $A = B$. This is shown in Fig. 9.8.

However, even if the two fuzzy variables are not strictly equal, that is, $\mu_A(x) \neq \mu_B(x)$, where $x \in X$, it is still possible to quantify, with $C_{eq}(A, B)$, how credible is the statement $A = B$. In fact, in this case, this coefficient estimates how much the two fuzzy variables overlap, by evaluating the ratio between their intersection area and their union area. Of course, in this case, this coefficient can reach the unitary value, by varying the relative position of the two fuzzy variables, only when the two membership functions have the same shape. This is shown in Figs. 9.9 and 9.10.

In Fig. 9.9, the two fuzzy variables A and B have the same shape for the membership functions, but it is $\mu_A(x) \neq \mu_B(x)$. It can be noted that

---

[1] Except for the limit case when both fuzzy variables degenerate into crisp variables. In this case, however, the application of this approach should be useless, being the ordering of two scalar values immediate.

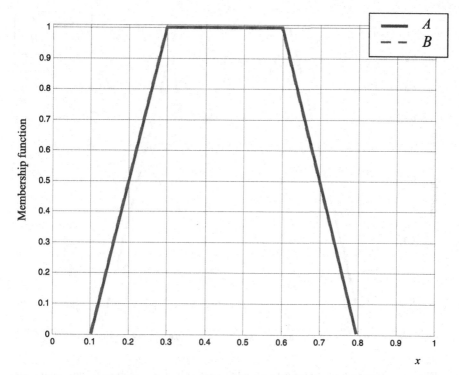

Fig. 9.8. Example of two fuzzy variables for which it is $\mu_A = \mu_B$. The intersection area coincides with the union area and $C_{eq}(A, B) = 1$.

these fuzzy variables are similar to those in Fig. 9.8, except for their relative position. In fact, in Fig. 9.8, the two considered fuzzy variables overlap totally, whereas in Fig. 9.9, they overlap only partially. Hence, in this last case, the intersection area is smaller than the union area, thus leading to a coefficient $C_{eq}(A, B)$ smaller than 1.

Let us now suppose to keep fuzzy variable A in its position, and move fuzzy variable B along the x axis. Then, the relative position of the two fuzzy variables varies, as well as their intersection area and their union area. When the two fuzzy variables are completely separated, the intersection area is zero, and $C_{eq}(A, B) = 0$. When the two fuzzy variables overlap totally, as in Fig. 9.8, the intersection area is equal to the union area, and $C_{eq}(A, B) = 1$. Therefore, by varying the relative position of the two considered fuzzy variables, $C_{eq}(A, B)$ varies in the range [0,1].

On the other hand, Fig. 9.10 shows two fuzzy variables with different shapes for their membership functions. In this case, whichever is the relative position of the two considered fuzzy variables, the intersection area is smaller than the union area, so that $C_{eq}(A, B)$ is always smaller than 1.

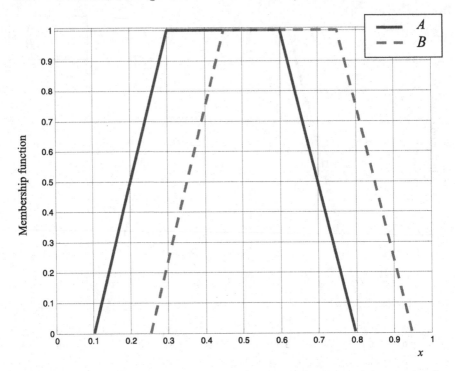

Fig. 9.9. Example of two fuzzy variables with the same shape of the membership functions. As it is $\mu_A \neq \mu_B$, then $C_{eq}(A, B) < 1$. However, by varying the relative position of A and B, $C_{eq}(A, B) = 1$ can be reached.

Hence, it can be concluded that, whichever are the considered fuzzy variables, coefficient $C_{eq}(A, B)$ can be used as the credibility coefficient of the statement $A = B$.

Under these assumptions, Eq. (9.3) becomes

$$C_{lo}(A, B) + C_{gr}(A, B) + C_{eq}(A, B) = 1$$

and can be interpreted as follows: Given to fuzzy variables A and B, the full credibility is given by the sum of the credibility that A is greater than B, that A is lower than B, and that A is equal to B. Therefore, these three credibility coefficients can be the basis for a decision rule.

9.2.2 Evaluation of the credibility coefficients

Let us now consider how the three credibility coefficients defined in the previous section can be evaluated.

The evaluation of $C_{eq}(A, B)$, defined by Eq. (9.6), is immediate and does not need any further explanation, because the union area and the intersection area have been already defined in Chapter 6. Hence, it is only necessary to

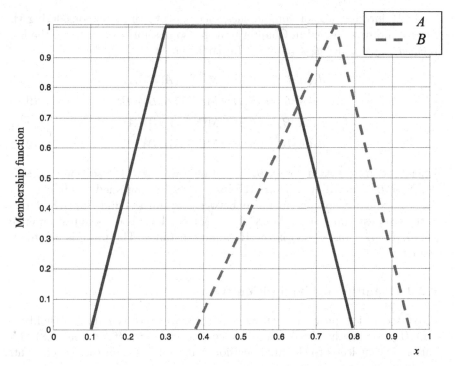

Fig. 9.10. Example of two fuzzy variables with a different shape of the membership functions. As it is $\mu_A \neq \mu_B$, then $C_{eq}(A, B) < 1$, whichever is their relative position.

examine Eqs. (9.4) and (9.5). In this respect, let us compare Figs. 9.5 and 9.7. By this comparison, it can be readily obtained that

$$d_1 + d_4 = d_A$$
$$d_2 + d_3 = d_B$$

where d_A and d_B are the Hamming distances defined by Kerre in its method; that is,

$$d_A = d(A, \mathrm{MAX}(A, B))$$
$$d_B = d(B, \mathrm{MAX}(A, B))$$

Therefore, it can be concluded that all three credibility coefficients C_{gr}, C_{lo}, and C_{eq} can be easily evaluated. In particular, Eqs. (9.4) and (9.5) can be rewritten as

$$C_{lo}(A, B) = \frac{d_A}{Un(A, B)} \tag{9.7}$$

$$C_{gr}(A, B) = \frac{d_B}{Un(A, B)} \tag{9.8}$$

Only for the sake of completeness, let us note that, by considering the fuzzy operators defined in Chapter 6, it is also possible to determine the four areas d_1, d_2, d_3, and d_4 separately. In particular, it is

$$d_1(A, B) = d((A \cap \text{MAX}^+(A, B)), A) \qquad (9.9)$$

$$d_2(A, B) = d((B \cap \text{MAX}^+(A, B)), B) \qquad (9.10)$$

$$d_3(A, B) = d((A \cap \text{MIN}^-(A, B)), A) \qquad (9.11)$$

$$d_4(A, B) = d((B \cap \text{MIN}^-(A, B)), B) \qquad (9.12)$$

where the intersection refers to the standard fuzzy intersection; variables $\text{MAX}(A, B)$ and $\text{MIN}(A, B)$ are the fuzzy variables obtained by applying the fuzzy-max and fuzzy-min operators between variables A and B; $\text{MAX}^+(A, B)$ is the greatest upper set of $\text{MAX}(A, B)$; $\text{MIN}^-(A, B)$ is the greatest lower set of $\text{MIN}(A, B)$; and $d(\bullet, \bullet)$ is the Hamming distance. The proof is left to the reader.

9.2.3 Definition of the decision rules

As stated in the previous sections, the comparison of two fuzzy variables A and B provides the three credibility coefficients (9.6), (9.7), and (9.8). The rationale that leads to the final decision is indeed based on these coefficients.
  The following can be stated.

- If $C_{gr}(A, B) > C_{lo}(A, B)$, then the final decision is that $A > B$. This decision is taken with a credibility factor equal to $C_{gr}(A, B)$.
- If $C_{lo}(A, B) > C_{gr}(A, B)$, then the final decision is that $A < B$. This decision is taken with a credibility factor equal to $C_{lo}(A, B)$.
- If $C_{gr}(A, B) = C_{lo}(A, B)$, then the final decision is that $A = B$. This decision is taken with a credibility factor equal to $C_{eq}(A, B)$.

Then, if the two fuzzy variables A and B in the example of Fig. 9.7 are reconsidered, it is

$C_{gr}(A, B) = 0.725$
$C_{lo}(A, B) = 0.032$
$C_{eq}(A, B) = 0.243$

Therefore, in this example, the final decision is that fuzzy variable A is greater than the fuzzy variable B, and this decision is taken with a credibility factor equal to 0.725.
  Let us consider that, when the final decision is that either $A > B$ or $A < B$, the credibility factor represents the degree of belief that $A > B$ or $A < B$, respectively. In these cases, the value assumed by the credibility factor is the greater, the lower is the intersection area between the two considered fuzzy variables A and B. In fact, it is obvious that, if the considered fuzzy variables do not overlap, the intersection area is nil, and the decision can be

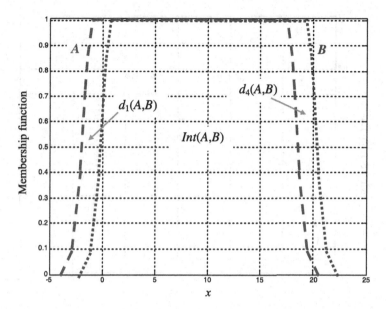

Fig. 9.11. Example of two fuzzy variables whose membership functions are mostly overlapping.

taken with full certainty, that is, with a unitary credibility factor. On the other hand, when the considered fuzzy variables overlap, the credibility factor associated with the final decision assumes a value that is lower than 1. Let us also consider that a decision can be taken with a credibility factor that is near zero. In this respect, let us consider Fig. 9.11. When the two fuzzy variables A and B shown in this figure are considered, it follows that both areas $d_2(A, B)$ and $d_3(A, B)$ are nil, and the intersection area prevails over the sum of the area $d_1(A, B)$ and $d_4(A, B)$. Therefore, it follows that $C_{gr}(A, B)$ is surely zero and $C_{eq}(A, B)$ is surely greater than $C_{lo}(A, B)$. In fact, for the considered example, it is

$$C_{gr}(A, B) = 0$$
$$C_{lo}(A, B) = 0.163$$
$$C_{eq}(A, B) = 0.837$$

According to the defined decision rules, the final decision is that fuzzy variable A is lower than fuzzy variable B, with a credibility factor equal to 0.163.

However, this example deserves further consideration. In fact, for this particular example, a doubt might arise whether to decide that fuzzy variable A is lower than fuzzy variable B, or to decide that fuzzy variable A is equal to fuzzy variable B. This doubt arises from the fact that the credibility coefficient

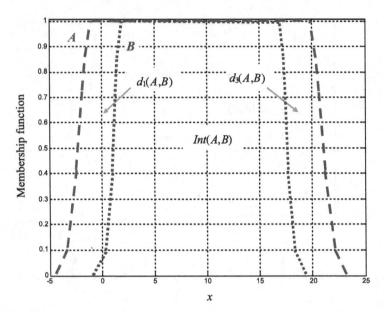

Fig. 9.12. Example of two fuzzy variables for which the credibility coefficients $C_{lo}(A,B)$ and $C_{gr}(A,B)$ take the same value.

$C_{eq}(A,B)$, which states the degree of equality, is greater than the credibility coefficient $C_{lo}(A,B)$, which states the minority.

Let us consider that, stating that the two fuzzy variables are equal, means declaring that there is not available evidence to assess that one variable is greater than the other one or vice versa. Instead, in the considered example, it is $C_{lo}(A,B) > C_{gr}(A,B)$. Hence, there is available evidence that fuzzy variable A is lower than fuzzy variable B, even if with a low credibility factor.

On the other hand, when the final decision is that $A = B$, the credibility factor associated with this statement assumes a quite different meaning. In fact, the final decision is that $A = B$ only in the very particular case when $C_{gr}(A,B) = C_{lo}(A,B)$. Therefore, we take this decision only when there is not enough available evidence to asses either that $A > B$ or that $A < B$.[2]

An example of this situation is shown in Fig. 9.12. For the two considered fuzzy variables A and B, it follows that both areas $d_2(A,B)$ and $d_4(A,B)$ are nil, and the two areas $d_1(A,B)$ and $d_3(A,B)$ are numerically equal. Therefore,

---

[2] Of course this situation, which can be defined in a very simple and immediate mathematical way, is almost never met in practice, because it is quite improbable that $C_{gr}(A,B)$ and $C_{lo}(A,B)$ take exactly the same numerical value. Therefore, in practical applications, this condition should be changed into the weaker condition: $|C_{gr}(A,B) - C_{lo}(A,B)| < \epsilon$, where ϵ is a suitable positive constant, chosen according to the characteristics of the particular considered application. For instance, ϵ could be a suitable fraction of the union area of A and B.

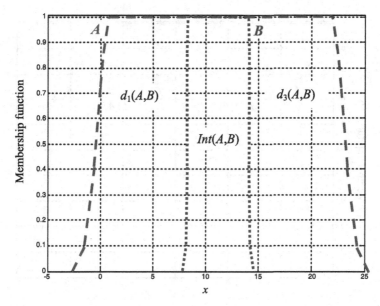

Fig. 9.13. Example of two fuzzy variables for which the credibility coefficients $C_{lo}(A, B)$ and $C_{gr}(A, B)$ take the same value. Even if $C_{eq}(A, B)$ takes a smaller value, the decision is that $A = B$.

it follows that $C_{gr}(A, B)$ is surely equal to $C_{lo}(A, B)$. In fact, for the considered example, it is

$$C_{gr}(A, B) = 0.15$$
$$C_{lo}(A, B) = 0.15$$

and there is no available evidence to assess which fuzzy variable is the greater. Then, according to the defined decision rules, the final decision is that fuzzy variable A is equal to fuzzy variable B. The credibility factor associated with this decision is given by coefficient $C_{eq}(A, B)$. It is

$$C_{eq}(A, B) = 0.7.$$

In this particular example, the intersection area prevails over both area $d_1(A, B)$ and area $d_3(A, B)$, so that coefficient $C_{eq}(A, B)$ is greater than both $C_{gr}(A, B)$ and $C_{lo}(A, B)$. Let us consider, however, that the same decision would also be taken in the case when coefficient $C_{eq}(A, B)$ was lower than the other two ones, as shown in Fig. 9.13.

Hence, it can be stated that the decision that two fuzzy variables are equal is taken each time the available evidence is not enough to establish an ordering of the two fuzzy variables themselves. The credibility factor associated with this statement represents a degree of belief in the statement itself: The greater is its value, the higher is the credibility associated with the decision; on the contrary, the smaller is this value, the higher is the amount of doubt

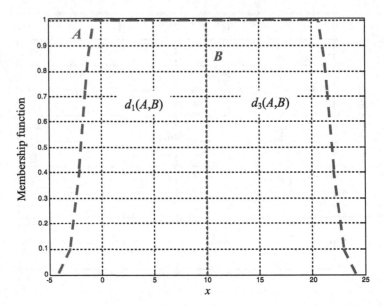

Fig. 9.14. Example where the decision that $A = B$ is taken with a zero credibility factor.

associated with the decision. In this respect, the decision that $A = B$ taken for the example of Fig. 9.12 is more certain than that taken for the example of Fig. 9.13, as can be intuitively stated, by considering that the two membership functions associated with the fuzzy variables in Fig. 9.12 are more similar than those in Fig. 9.13. In the limit case where fuzzy variable B degenerates into a scalar value, as shown in Fig. 9.14, the decision is taken with a zero credibility factor. This decision could seem illogical, but no evidence exists in support to a different decision.

9.2.4 Properties of the credibility coefficients

Let us now consider some properties of the credibility coefficients defined in the previous section. As stated, it is

$$C_{gr}(A, B) + C_{lo}(A, B) + C_{eq}(A, B) = 1$$

The above relationship must of course be independent on the order of the two considered variables; in other words, if A and B are interchanged, it is again

$$C_{gr}(B, A) + C_{lo}(B, A) + C_{eq}(B, A) = 1$$

Moreover, it can be intuitively expected that

$$C_{eq}(A, B) = C_{eq}(B, A) \tag{9.13}$$
$$C_{gr}(A, B) = C_{lo}(B, A) \tag{9.14}$$
$$C_{lo}(A, B) = C_{gr}(B, A) \tag{9.15}$$

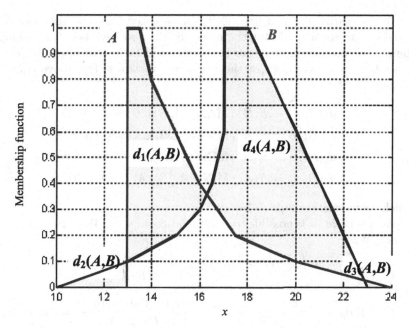

Fig. 9.15. Areas d_1, d_2, d_3, and d_4 for the fuzzy variables A and B. The variables are the same as those of Fig. 9.7, but their names have been interchanged.

Relationships (9.13), (9.14), and (9.15) can be readily proved, by considering that, by definition, the following equalities are valid:

$$\text{Int}(A, B) = \text{Int}(B, A) \tag{9.16}$$
$$Un(A, B) = Un(B, A) \tag{9.17}$$
$$d_1(A, B) = d_2(B, A) \tag{9.18}$$
$$d_2(A, B) = d_1(B, A) \tag{9.19}$$
$$d_3(A, B) = d_4(B, A) \tag{9.20}$$
$$d_4(A, B) = d_3(B, A) \tag{9.21}$$

The proof of Eqs. (9.16) and (9.17) is immediate. To prove relationships from Eq. (9.18) to (9.21), let us consider Figs. 9.7 and 9.15. In both figures, the four areas d_1, d_2, d_3, and d_4 are considered, as determined by applying Eqs. (9.9)–(9.12), here reported again for the sake of clearness:

$$d_1(A, B) = d((A \cap \text{MAX}^+(A, B)), A)$$
$$d_2(A, B) = d((B \cap \text{MAX}^+(A, B)), B)$$
$$d_3(A, B) = d((A \cap \text{MIN}^-(A, B)), A)$$
$$d_4(A, B) = d((B \cap \text{MIN}^-(A, B)), B)$$

By comparing the two figures, it can be noted that the same two fuzzy variables are considered, but the names of the two have been interchanged. Hence, by considering that the fuzzy-max and fuzzy-min operators are independent from the order of the two fuzzy variables, the above equations become, respectively:

$$d_1(B, A) = d((B \cap \mathrm{MAX}^+(A, B)), B) = d_2(A, B)$$
$$d_2(B, A) = d((A \cap \mathrm{MAX}^+(A, B)), A) = d_1(A, B)$$
$$d_3(B, A) = d((B \cap \mathrm{MIN}^-(A, B)), B) = d_4(A, B)$$
$$d_4(B, A) = d((A \cap \mathrm{MIN}^-(A, B)), A) = d_3(A, B)$$

which prove relationships (9.18)–(9.21).

At this point, it is possible to prove relationships (9.13), (9.14), and (9.15). In fact, it is

$$C_{eq}(A, B) = \frac{Int(A, B)}{Un(A, B)} = \frac{Int(B, A)}{Un(B, A)} = C_{eq}(B, A)$$

$$C_{gr}(A, B) = \frac{d_2(A, B) + d_3(A, B)}{Un(A, B)} = \frac{d_1(B, A) + d_4(B, A)}{Un(B, A)} = C_{lo}(B, A)$$

$$C_{lo}(A, B) = \frac{d_1(A, B) + d_4(A, B)}{Un(A, B)} = \frac{d_2(B, A) + d_3(B, A)}{Un(B, A)} = C_{gr}(B, A)$$

9.2.5 Analysis of the credibility coefficients

It is interesting to see how the three credibility coefficients vary as the relative position of two given fuzzy variables varies.

Let us consider Fig. 9.16. Let us suppose that fuzzy variable A is kept in its original position, whereas fuzzy variable B varies its position along the x axis. Let us select $b_1^{\alpha=0}$ (see Fig. 9.16), that is, the left edge of the α-cut at level $\alpha = 0$, as an indication of the position of B^3.

Let us now vary $b_1^{\alpha=0}$ from 0 to 27 and compare the fuzzy variables A and B for each relative position. Values of $b_1^{\alpha=0}$ below 0 and above 27 do not provide any further information, because in both cases, the two fuzzy variables are completely separated and do not overlap; hence, these values are not considered. Of course, as $b_1^{\alpha=0}$ varies, the four areas shown in Fig. 9.16 modify, and the credibility coefficients associated with the statements $A > B$, $A < B$, and $A = B$ take different values.

The values obtained for the three credibility coefficients as a function of the relative position of the two fuzzy variables are plotted in Figs. 9.17, 9.18, and 9.19.

---

[3] The choice of this point is totally arbitrary. If different points are considered, the same results are obtained.

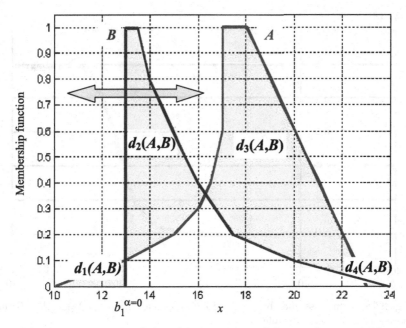

Fig. 9.16. Variation of the relative position of the two fuzzy variables.

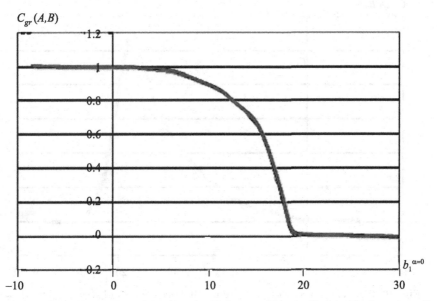

Fig. 9.17. Values assumed by the credibility coefficient $C_{gr}(A, B)$ that $A > B$ versus the position of fuzzy variable B.

$C_{lo}(A,B)$

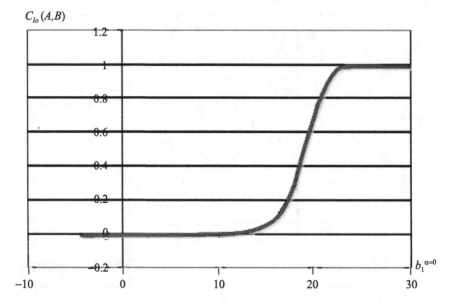

Fig. 9.18. Values assumed by the credibility coefficient $C_{lo}(A, B)$ that $A < B$ versus the position of fuzzy variable B.

$C_{eq}(A,B)$

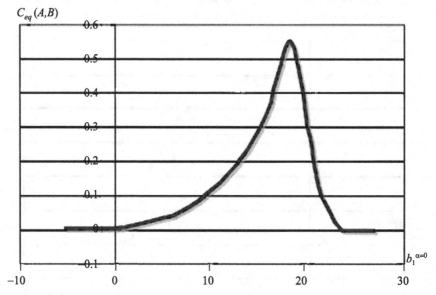

Fig. 9.19. Values assumed by the credibility coefficient $C_{eq}(A, B)$ that $A = B$ versus the position of fuzzy variable B.

Figure 9.17 shows the variation of coefficient $C_{gr}(A, B)$, which represents the degree of belief that A is greater than B. It can be noted that this coefficient starts from a unitary value, when $b_1^{\alpha=0}$ is zero (and fuzzy variable A is certainly greater than fuzzy variable B), and then decreases monotonically to zero as fuzzy variable B moves to the right and $b_1^{\alpha=0}$ increases.

Figure 9.18 shows the variation of coefficient $C_{lo}(A, B)$, which represents the degree of belief that A is lower than B. It can be noted that this coefficient starts from zero, when $b_1^{\alpha=0}$ is zero (and fuzzy variable A is certainly greater than fuzzy variable B), and then increases monotonically to one as fuzzy variable B moves to the right and $b_1^{\alpha=0}$ increases.

Figure 9.19 shows the variation of coefficient $C_{eq}(A, B)$, which represents the degree of belief that A is equal to B. It can be noted that this coefficient starts from zero, when $b_1^{\alpha=0}$ is zero and fuzzy variable A is certainly greater than fuzzy variable B; then it increases as fuzzy variable B starts moving to the right and overlapping with fuzzy variable A, and reaches its maximum value when the relative position of the two fuzzy variables leads to the maximum overlapping between them. $C_{eq}(A, B)$ then decreases to zero again as fuzzy variable B keeps moving and, finally, becomes certainly lower than fuzzy variable A.

Hence, coefficient $C_{eq}(A, B)$ is described by a convex function. In the example described in Fig. 9.19, the maximum of this function takes value 0.512. In general, the maximum of this function is a value lower than 1, and the unitary value is reached only in the particular situation when the two fuzzy variables have exactly the same shape of the membership functions. In fact, only in this case, a relative position can be found for which the two fuzzy variables are fully overlapping, the intersection area and the union area are the same, and coefficient $C_{eq}(A, B) = 1$.

By comparing the three credibility coefficients, it is possible to make the final decision and to evaluate its credibility factor. Figure 9.20 shows the credibility factors as the relative position of the two fuzzy variables varies. Three different zones can be considered. In the first zone, when $b_1^{\alpha=0}$ is lower than 19.6, the final decision is that fuzzy variable A is greater than fuzzy variable B. In the second zone, when $b_1^{\alpha=0} = 0$ is greater than 19.6, the final decision is that fuzzy variable A is lower than fuzzy variable B. In the third and last zone, which corresponds to the single value $b_1^{\alpha=0} = 19.6$, the final decision is that fuzzy variable A is equal to fuzzy variable B. In fact, in this particular situation, the two credibility coefficients $C_{gr}(A, B)$ and $C_{lo}(A, B)$ coincide.

Let us also note that the functions plotted in Figs. 9.17 and 9.18 can be regarded as membership functions. These membership functions describe the degrees of belief associated with the statements that A is greater than B and A is lower than B, respectively, as the relative position of the two considered fuzzy variables varies.

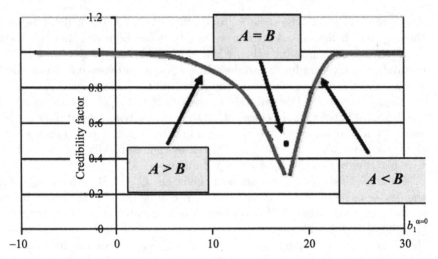

Fig. 9.20. Credibility factors at the different possible relative positions of the considered fuzzy variables.

Hence, it can be stated that the result of the different comparisons between two fuzzy variables provides a fuzzy variable again. This is completely compatible with the followed fuzzy approach.

Moreover, the obtained membership functions can be employed in fuzzy inference, whenever more complex decisional algorithms should be elaborated.

10

List of Symbols

α	degree of membership
A_-	greatest lower set of a generic fuzzy variable A
A_+	greatest upper set of a generic fuzzy variable A
A_α	α-cut of a generic fuzzy variable A
Bel	belief function
$C_{eq}()$	equal credibility coefficient
$C_{gr}()$	greater credibility coefficient
$C_{lo}()$	lower credibility coefficient
$d()$	Hamming distance
$Dou()$	doubt function
$f()$	generic function
$f'()$	first derivative of function $f()$
$h()$	aggregation operator
$Int()$	intersection area
m	basic probability assignment function
$M()$	averaging operator
$\mu(x)$	membership function of a fuzzy variable
$\max()$	standard fuzzy union
$MAX()$	fuzzy-max operator
$\max\{\}$	max function
$\min()$	standard fuzzy intersection
$MIN()$	fuzzy-min operator
$\min\{\}$	min function
$Nec()$	necessity function
$p(x)$	probability distribution function
$Pl()$	plausibility function
$Pos()$	possibility function
$Pro()$	probability function
ρ	correlation coefficient

$r(x)$	possibility distribution function
σ	standard deviation
$S()$	t-conorm
$T()$	t-norm
$Un()$	union area
X	universal set
Y_A	Yager area of a generic fuzzy variable A

References

[CH92]   S. J. Chen, C. L. Hwang: Fuzzy Multiple Attribute Decision Making: Methods and Applications, Springer-Verlag, New York (1992)

[DP93]   Dubois, Prade: Fuzzy sets and probability: misunderstanding, bridges and gaps, Proc. Second IEEE International Conference on Fuzzy Systems, San Francisco, pp. 1059–1068, (1993)

[FGS04]   A. Ferrero, R. Gamba, S. Salicone: A method based on random-fuzzy variables for on-line estimation of the measurement uncertainty of DSP-based instruments, IEEE Transactions on Instrumentation and Measurement, pp. 1362–1369, (October 2004)

[FS03]   A. Ferrero, S. Salicone: An innovative approach to the determination of uncertainty in measurement based on fuzzy variables, IEEE Transactions on Instrumentation and Measurement, vol. 52, no. 4, pp. 1174–1181, (2003)

[FS04]   A. Ferrero, S. Salicone: The random-fuzzy variables: a new approach for the expression of uncertainty in measurement, IEEE Transactions on Instrumentation and Measurement, pp. 1370–1377, (October 2004)

[FS05a]   A. Ferrero, S. Salicone: The use of random-fuzzy variables for the implementation of decision rules in the presence of measurement uncertainty, IEEE Transactions on Instrumentation and Measurement, pp. 1482–1488, (August 2005)

[FS05b]   A. Ferrero, S. Salicone: A comparative analysis of the statistical and random-fuzzy approaches in the expression of uncertainty in measurement, IEEE Transactions on Instrumentation and Measurement, pp. 1475–1481, (August 2005)

[FS05c]   A. Ferrero, S. Salicone: The Theory of Evidence for the expression of uncertainty in measurement—La théorie de l'évidence pour l'expression de l'incertitude dans les mesures, Proc. International Metrology Congress, Lyon, France, June 20–23, (2005)

[KG91]   A. Kaufmann, M. M. Gupta: Introduction to Fuzzy Arithmetic, Theory and Applications, Van Nostrand Reinhold, New York, (1991)

[KH94]   G. J. Klir, D. Harmanec: On modal logic interpretation of possibility theory, International Journal of Uncertainty, Fuzziness and Knowledge-Based Systems, vol. 2, pp. 237–245, (1994)

[KY95]   G. J. Klir, B. Yuan: Fuzzy Sets and Fuzzy Logic. Theory and Applications, Prentice Hall PTR, Englewood Cliffs, NJ (1995)

[ISO04]   IEC-ISO Guide to the Expression of Uncertainty in Measurement, Supplement 1, Numerical Methods for the Propagation of Distributions

[ISO93]   IEC-ISO Guide to the Expression of Uncertainty in Measurement (1993)

[M03]   L. Mari: Epistemology of measurement, Measurement, vol. 34, pp. 17–30 (2003)

[MBFH00]   G. Mauris, L. Berrah, L. Foulloy, A. Haurat: Fuzzy handling of measurement errors in instrumentation, IEEE Transactions on Instrumentation and Measurement, vol. 49, no. 1, pp. 89–93, (2000)

[MLF01]   G. Mauris, V. Lasserre, L. Foulloy: A fuzzy approach for the expression of uncertainty in measurement, Measurement 29, pp. 165–177, (2001)

[MLF97]   G. Mauris, V. Lasserre, L. Foulloy: A simple probability-possibility transformation for measurement error representation: a truncated triangular transformation, in World Congress of International Fuzzy Systems Assoc., IFSA, Prague, Czech Republic, pp. 476–481, (1997)

[P91]   A. Papoulis: Probability, Random Variables and Stochastic Processes, McGraw-Hill, New York (1991)

[R89]   E. H. Ruspini: The semantics of vague knowledge, Revue Internationale de Systemique, vol. 3, pp. 387–420 (1989)

[S76]   Glenn Schafer: A Mathematical Theory of Evidence, Princeton University Press, Princeton, NJ (1996)

[SS63]   B. Schweizer, A. Sklar: Associative functions and abstract semigroups, Publ. Math. Debrecen, no. 10, pp. 69–81 (1963)

[UW03]   M. Urbansky, J. Wasowski: Fuzzy approach to the theory of measurement inexactness, Measurement 34, pp. 67–74, (2003)

[Z65]   L. A. Zadeh: Fuzzy sets, Information and Control, vol. 8, pp. 338–353 (1965)

[Z73]   L. A. Zadeh: Outline of a new approach to the analysis of complex systems and decision processes, IEEE Transactions on Systems, Man and Cybernetics vol. SMC-2, pp. 28–44 (1973)

[Z75]   L. A. Zadeh: Fuzzy logic and approximate reasoning, Synthese 30, pp. 407–428 (1975)

[Z78]   L. A. Zadeh: Fuzzy sets as a basis for a theory of possibility, Fuzzy Sets and Systems, vol. 1, pp. 3–28 (1978)

Index